Game Dynamics

Oliver Korn · Newton Lee
Editors

Game Dynamics

Best Practices in Procedural and Dynamic Game Content Generation

Editors
Oliver Korn
Offenburg University
Offenburg
Germany

Newton Lee
Newton Lee Laboratories, LLC
Tujunga, CA
USA

ISBN 978-3-319-85059-7 ISBN 978-3-319-53088-8 (eBook)
DOI 10.1007/978-3-319-53088-8

© Springer International Publishing AG 2017
Softcover reprint of the hardcover 1st edition 2017
This work is subject to copyright. All rights are reserved by the Publisher, whether the whole or part of the material is concerned, specifically the rights of translation, reprinting, reuse of illustrations, recitation, broadcasting, reproduction on microfilms or in any other physical way, and transmission or information storage and retrieval, electronic adaptation, computer software, or by similar or dissimilar methodology now known or hereafter developed.
The use of general descriptive names, registered names, trademarks, service marks, etc. in this publication does not imply, even in the absence of a specific statement, that such names are exempt from the relevant protective laws and regulations and therefore free for general use.
The publisher, the authors and the editors are safe to assume that the advice and information in this book are believed to be true and accurate at the date of publication. Neither the publisher nor the authors or the editors give a warranty, express or implied, with respect to the material contained herein or for any errors or omissions that may have been made. The publisher remains neutral with regard to jurisdictional claims in published maps and institutional affiliations.

Printed on acid-free paper

This Springer imprint is published by Springer Nature
The registered company is Springer International Publishing AG
The registered company address is: Gewerbestrasse 11, 6330 Cham, Switzerland

Contents

A Very Short History of Dynamic and Procedural Content Generation 1
Michael Blatz and Oliver Korn

Procedural Content Generation in the Game Industry 15
Alba Amato

Design, Dynamics, Experience (DDE): An Advancement of the MDA Framework for Game Design 27
Wolfgang Walk, Daniel Görlich and Mark Barrett

Procedural Synthesis of Gunshot Sounds Based on Physically Motivated Models 47
Hüseyin Hacıhabiboğlu

Dynamic Player Pairing: Quantifying the Effects of Competitive Versus Cooperative Attitudes 71
Gerry Chan, Anthony Whitehead and Avi Parush

FaceMaker—A Procedural Face Generator to Foster Character Design Research 95
Valentin Schwind, Katrin Wolf and Niels Henze

A Primer on Procedural Character Generation for Games and Real-Time Applications 115
Yanko Oliveira

Procedural Terrain Generation. A Case Study from the Game Industry 133
Jakob Schaal

Procedural Adventure Generation: The Quest of Meeting Shifting Design Goals with Flexible Algorithms 151
Manuel Kerssemakers

Index 175

About the Editors

About the Editors

Oliver Korn is a full professor for Human Computer Interaction (HCI) at Offenburg University in Germany. He also is a certified project manager (German Chamber of Commerce, DIHK), professional member of the ACM and the IEEE, and an evaluator for the European Commission.

After completing his master's focusing on Computational Linguistics, he worked at the Fraunhofer Institute for Industrial Engineering (IAO) and the Stuttgart Media University (HdM). He focused on HCI, especially simulations, gaming, and gamification. In 2003, he co-founded KORION, a Fraunhofer spin-off developing simulations and games. As CEO, he was in charge of several national research projects and gained experience in countless industrial software development projects, both in the area of entertainment and business intelligence. In 2014, he received his Ph.D. in computer science at the SimTech Excellence Cluster of the University of Stuttgart. His research is published in numerous international publications.

He has been a game enthusiast since the times of the ZX81 and loves games with dynamic and procedural contents like *Civilization* or *Master of Magic*. Today, he works on the convergence of digital technology and real life, focusing on affective computing, gamification, augmented work, and learning. Korn's overall vision is integrating gameful design in education, health and work processes, augmenting and enriching everyday activities. In the area of affective computing, he aims to assess emotional states to improve computational context awareness.

Newton Lee is CEO of Newton Lee Laboratories LLC, president of the Institute for Education, Research, and Scholarships (IFERS), adjunct professor at Woodbury University School of Media, Culture & Design (MCD), editor in chief of Association for Computing Machinery (ACM) Computers in Entertainment, and education and media advisor for the United States Transhumanist Party.

Previously, Lee was a computer scientist at AT&T Bell Laboratories, research staff member at the Institute for Defense Analyses, and senior producer and engineer at The Walt Disney Company where he led the in-house Games Group in developing over 100 online games for Disney.com and DisneyBlast.com. He was the founder of Disney Online Technology Forum, creator of Bell Labs' first-ever commercial AI tool, and inventor of the world's first annotated multimedia OPAC for the US National Agricultural Library.

Lee graduated Summa Cum Laude from Virginia Tech with a B.S. and M.S. degree in computer science, and he earned a perfect GPA from Vincennes University with an A.S. degree in electrical engineering and an honorary doctorate in computer science. He has been honored with a Michigan Leading Edge Technologies Award, two community development awards from the California Junior Chamber of Commerce, and four volunteer project leadership awards from The Walt Disney Company.

An Introduction to Dynamic and Procedural Content Generation

Yet all joy wants eternity–
–Wants deep, wants deep eternity.

Friedrich Nietzsche, *Thus Spoke Zarathustra*, translation by Adrian Del Caro

This famous quote from Zarathustra's Roundelay addresses the central human desire that fun and play may last eternally. So were you among those who enjoyed the endless variety in classic games such as *Masters of Magic* or *Civilization*? Did you spend night after night creating new creatures in *Spore* or investing "just one more hour" to get that unique epic gun with just the right attributes in *Borderlands*? If you answered yes, you got caught in the potentially eternal delights of procedurally generated, dynamic game content. Furthermore, you probably are about the same age as the editors. If you are the younger generation, games like *No Man's Sky* or *Renowned Explorers* represent modern iterations of the procedural method.

Dynamic game content may not achieve the cinematic quality of some of the peak moments in more linear games such as *Call of Duty*, but it surely creates a special feeling during play that things are endless and that fun and play can potentially last forever.

The editors both have had extensive experience with "fun and games"—for 10 years in the game industry, Oliver Korn was at the core of creating gameful experiences—both for entertainment games and gamified real-world applications. Designing experiences that constantly create hedonic moments inevitably led him to procedural content generation. As a professor for Human Computer Interaction, he now aims to analyze and describe the formulae that maintain the flow. At The Walt Disney Company, Newton Lee worked on the same topic—during a 10-year period, he supervised the design and development of over 100 online games—some of which were discussed in his 2012 book *Disney Stories: Getting to Digital*. Disney artists are known to meticulously craft every animation and design each background in a computer game according to an extensive set of style guides. To measure the success of a game, replay value (aka replayability) is just as important as stunning visual elements, compelling story, and game mechanics. Dynamic and procedural content generation with a touch of artificial intelligence go a long way toward achieving high replayability.

In this unique book, we present several aspects of procedural and dynamic methods related to game design, game studies, and research. It is oriented toward industry best practices but it also covers relevant backgrounds, models, and methods, providing deep insights into the science behind procedural and dynamic approaches.

The chapters are not building upon each other. Instead, it is the book's common theme that connects all the chapters. In addition, a subject index shows recurring concepts, persons, and of course games. If you look at the author profiles, you will find two main groups: game designers with an affinity toward academia, and academics with an affinity toward game design. However, this distinction becomes blurry for the book editors and some contributing authors. This fuzzy border is a good thing: It shows that the worlds of academia and industry in gaming are overlapping. We believe that designing games means creating an interactive form of art that draws heavily on computer science. With procedural and dynamic gameplay, this proximity increases, as the importance of coding and mathematics increases with each element created algorithmically instead of manually by an artist.

Most of you, dear readers, will probably jump right to the chapter that you find most interesting. Nevertheless, it was our task as editors to bring the chapters into a logical sequence. We decided to apply the classic ordering process starting with the basics such as history and models (Korn and Blatz; Walk et al.), continuing with a survey on the state of the art (Amato), moving to methods (Chan et al.; Hacıhabiboğlu; Schwind et al.), and finally to best practices (Oliveira; Schaal; Kersemakers). The classification of FaceMaker by Schwind et al. as a method and procedural character generation by Oliveira as a best practice shows that the borders are blurry once again. However, if you appreciate procedural approaches, this little bit of randomness will surely not frighten you.

We wanted to make this book not only an intellectual but also a visual pleasure—integrating screenshots from games as well as visualizations of models and methods. Of course, the best practice articles have advantages in this respect, as visuals require works that are closer to completion. Due to the applied method of sequencing the chapters, this does result in the book becoming increasingly colorful with each chapter and each page, just like the complexity of a procedural algorithm's output typically grows with each iteration.

We wish you a vivid and inspiring reading experience, and we hope that this book may contribute to bridging the gap between computer science and game art, inspiring you to create more and better games or to invest more research in this fascinating field.

Oliver Korn
Newton Lee

A Very Short History of Dynamic and Procedural Content Generation

Michael Blatz and Oliver Korn

Abstract This chapter portrays the historical and mathematical background of dynamic and procedural content generation (PCG). We portray and compare various PCG methods and analyze which mathematical approach is suited for typical applications in game design. In the next step, a structural overview of games applying PCG as well as types of PCG is presented. As abundant PCG content can be overwhelming, we discuss context-aware adaptation as a way to adapt the challenge to individual players' requirements. Finally, we take a brief look at the future of PCG.

1 Introduction and Motivation

Game contents like terrain, characters, items and even story elements can be developed in two radically different ways: either by manual content creation: Artists develop graphics, writers create stories, etc., or by procedural content generation (PCG). With PCG, algorithms generate the content procedurally, either before a level starts or even continuously during runtime.

However, procedural structures have been there long before the invention of computers and even long before humankind. Many elemental and natural structures both on very large scales (galaxies, island distribution, coastlines, etc.) and on smaller scales (reef structures, some vegetables, bacteria growth, etc.) show that the

---

M. Blatz
KORION GmbH, Ludwigsburg, Germany
e-mail: michael.blatz@korion.de

O. Korn (✉)
Offenburg University, Offenburg, Germany
e-mail: oliver.korn@acm.org

© Springer International Publishing AG 2017
O. Korn and N. Lee (eds.), *Game Dynamics*,
DOI 10.1007/978-3-319-53088-8_1

Fig. 1 Romanesco is a good example of how iterative processes can create aesthetically pleasing self-similar structures

structures generated by the iteration of simple rules are in fact universal. Even a common vegetable like Romanesco shows these "self-similar" structures, and thus the graphical effects of PCG at work (Fig. 1).

While PCG has always played a role in game development, it is especially influential today. Platforms like Steam, App Store or Google Play allow an easier distribution of games without publishers and distribution media. In consequence, the video game sector, which was already very industrialized, is reliving its early days: Independent developers with small teams create thousands of creative applications. These development teams typically have neither the number of skilled employees nor the budget to create manually substantial amounts of game content. Thus, either the "indies" create very small games—or they delve into PCG. Applying PCG usually results in saving artists' capacity by modifying or even creating assets with the use of algorithms, typically using a variety of randomized parameters. On the graphical level, this results in different looks of rocks, trees, weapons or even whole buildings—eventually, the whole game world could be generated procedurally (Hosking 2013).

A large scale of potential game design patterns accompanies these possibilities. Only a few years back, it was primarily the technical possibilities, which limited how games looked and felt like. Today the simple "pixel look" of *Minecraft* (Persson 2011) is used deliberately as a stylistic device and exists independently besides highly detailed, almost photo-realistic graphics. The current spectrum in computer game designs reaches from pixel graphics as in the era of the Commodore 64 (Jones 2013) to high-definition, 3D graphics. The spectrum of stories reaches from simple jump and run ("rescue the princess") to role-playing games offering fascinating and complex experiences even for players familiar with top-notch movies and series.

However, even fascinating stories or excellent graphics are no guarantee for appreciation. A good example is the *Call of Duty* series, which undoubtedly is produced with great financial and artistic effort. Nevertheless, in player forums the games are often criticized for their tube-like linear level maps, which leave players little freedom of choice. Such restrictions are characteristic flaws of manually generated stories and levels, making *Call of Duty* a typical representative of state-of-the-art manual content generation. At the same time the critics should realize that the richness of detail and the cinematic quality of some of the "epic" moments in this series can only be achieved through a high level of control, inevitably at the cost of the players' freedom (Yannakakis and Togelius 2011). When aiming at such cinematic experiences, much budget is required for the development of high-quality assets. Therefore, the interest in PCG grows even in the traditional game industry.

In Sect. 2 of this chapter, we portray the historical and mathematical background of generating content dynamically and procedurally and compare various methods. We mainly focus on graphical aspects, the area where PCG originated. In Sect. 3, we present a structural overview of games applying PCG as well as different types of PCG. We discuss a method to adapt the challenge to individual players' requirements. Finally, we take a brief look at the future of PCG.

2 PCG Methods

For a better understanding of how PCG can be utilized in games, we portray the most influential methods to create procedural content. Algorithms for procedural content are often applied in computer games, yet there are others mainly focusing on artificial intelligence (e.g., AI Director) or changes of state (e.g., Markov chains). Therefore, the particular use case defines the implementation of specific PCG methods.

The most common use of PCG is probably generating terrain or landscapes (Shaker et al. 2016). However, the underlying algorithms can be adapted for other use cases. Typically, noise functions are used to create the content, but principally other solutions like fractal-based algorithms could also generate the required self-similar structures. To provide an overview of the algorithmic potentials, we briefly describe common PCG methods and their historical background.

2.1 *Brown Fractals and Mandelbrot Sets*

The Brownian motion is an important starting point in the history of procedural algorithms. Its discovery goes back to the botanist Robert Brown (1773–1858), who observed the random motion of pollen grain particles within water. This motion is the result of non-visible molecules in liquids and gasses that move with random velocities, in random directions, colliding with the pollen grains. As a natural phenomenon, it is one of the simplest observable stochastic processes.

Almost a century later, the mathematician Norbert Wiener (1894–1964) was inspired by this phenomenon and together with Godfrey Harold "G. H." Hardy tried to map it in an algorithm. Later the French physicist and Nobel-prize winner Jean Perrin (1870–1942) had the idea of describing the Brownian motion as continuous-time stochastic process by using non-differentiable curves (Mandelbrot 1982)—a method called "Wiener process" (visualized in Fig. 2) in honor of Wiener.

Benoit Mandelbrot (1924–2010) later frequently described the Wiener process as one of the keystones of his work. Mandelbrot used the Brownian motion to develop what he calls a "self-avoiding random walk" (Fig. 3). Clearly, this implementation is well suited to generate forms that are similar to land masses or jagged coastlines.

Another algorithmic approach is fractals, defined as a mathematical set that exhibits a repeating pattern that displays at every scale (Boeing 2016). From a geometrical point of view, fractals are self-similar structures (Fig. 4) based on recursive functions with potentially infinite depth. The very common "Julia sets" or "Mandelbrot sets" are such fractal structures, the latter one named after the discoverer Mandelbrot. It is visually striking, how several plants (Fig. 1) but also macroscopic structures like coastal lines of reefs incorporate structures similar to fractals. This phenomenon indicates that such natural structures can be algorithmically described using fractals. Mandelbrot even claimed that all forms of nature like trees, mountains, coastlines or clouds can only adequately be described using fractals (Ziegler 2013).

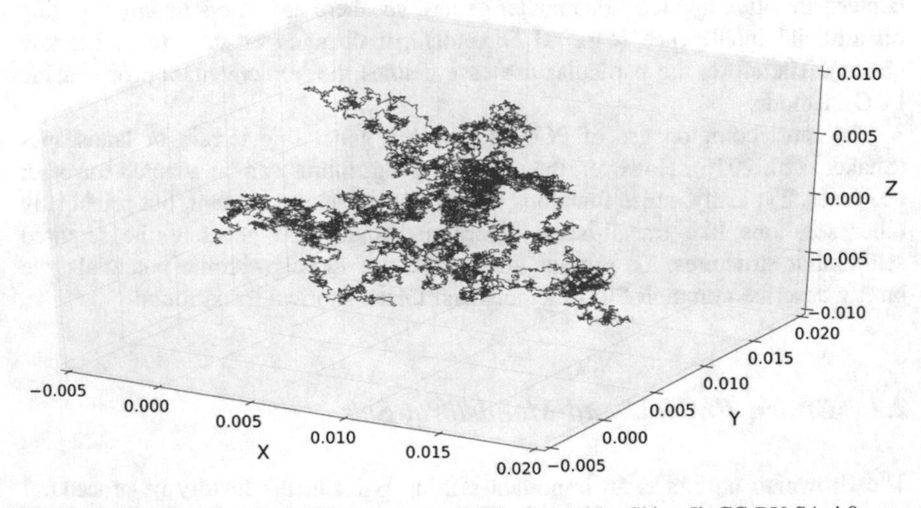

Fig. 2 Realization of a single Wiener process in 3D. Created by Shiyu Ji, CC BY-SA 4.0

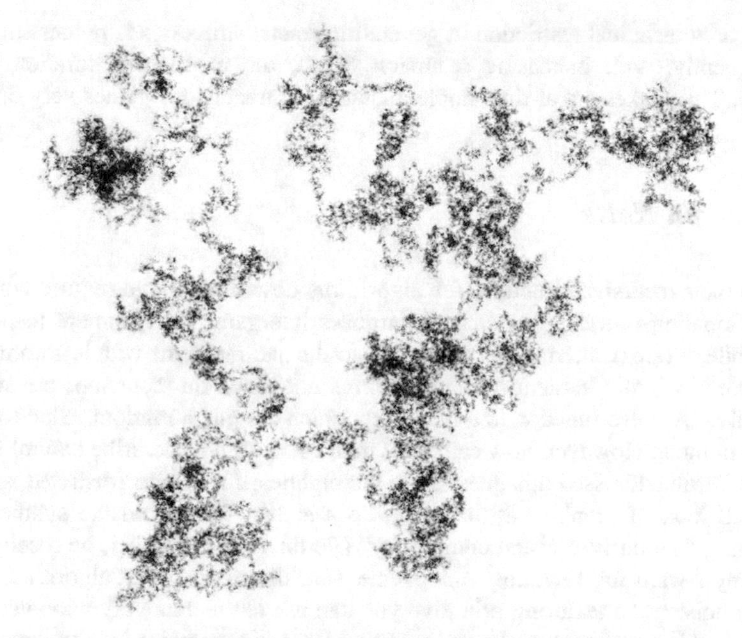

Fig. 3 Self-avoiding random walk in two dimensions. The *darker* the region, the more it has been visited. Created with MATLAB by Purpy Pupple, CC BY-SA 3.0

Fig. 4 Mandelbrot set probably is the most popular geometrical fractal. Its self-similar iterative structure is found in both small and large structures, e.g., vegetables and reefs. The image was generated with the easy fractal generator

The only technical restriction in generating fractal structures is processing time. Consequently, with increasing recursion depth, the processing duration grows quickly. This makes a real-time implementation of fractals for games very difficult.

2.2 *Perlin Noise*

Due to their recursive structure, the algorithms described above require considerable computing power. For practical purposes like game development (especially for mobile devices), alternatives that can describe natural forms with less computing effort are required. Gradient noise algorithms not based on recursions are such an alternative. A noise function is a mapping, which assigns a random value to every natural number. However, how can noise images be used to describe natural forms?

Ken Perlin addressed this question in the eighties. As he was frustrated with the pixelated look of graphics at that time, he searched for an "image synthesizer," generating "naturalistic visual complexity" (Perlin 1985). In 1997, he received the Academy Award for Technical Achievement for discovering that algorithm. Perlin defined noise as "a texturing primitive you can use to create a very wide variety of natural looking textures" and explained that "combining noise into various mathematical expressions produces procedural texture" (Perlin 1999). This texture can be applied to any object, so Perlin noise can be used in 2D as well as in 3D (Fig. 5).

Another advantage of gradient noise functions is that, even if the distribution is randomly generated, every configuration can be saved and reapplied so that the same conditions will always produce the same results. Perlin used frequency and amplitude to describe the function. The texture is a single-channel image (usually gray scale) where pixel values represent frequencies between −1 and +1. A high frequency (up to 1) means smaller details, resulting in many small points in a specific area, while a low frequency results in mostly large points. The noise function's amplitude describes the color values' "height" between completely black (0) and completely white (255). The implementation is described in a four-step process (Perlin 1999):

1. Given an input point
2. For each of its neighboring grid points: pick a "pseudo-random" gradient vector.
3. Compute linear function (dot product)
4. Take weighted sum, using ease curves.

Later a revised version called "Simplex Noise" was developed (Olano et al. 2003), as the first implementation only performed well in up to three dimensions. Simplex noise does not use a grid based on quads, but a grid with equal-sided triangles, reducing the number of neighboring points and therefore computing time. Contrary to Perlin noise with the complexity of $O(2n)$, simplex noise only requires $2(n^2)$, using only $n + 1$ vertices instead of $2n$. This is most noticeable when working in dimensions with $n > 4$.

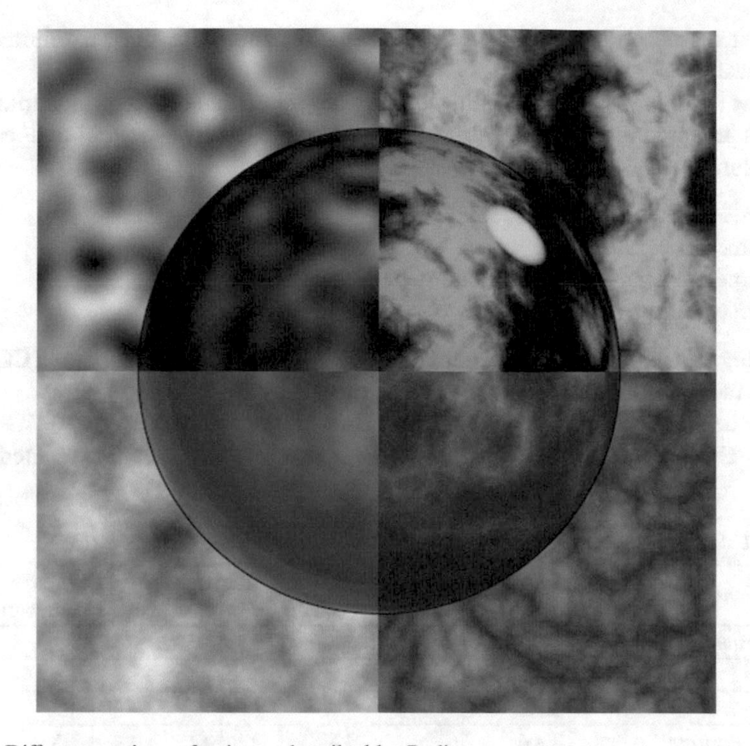

Fig. 5 Different versions of noise as described by Perlin

3 State-of-the-Art

We already introduced PCG as a versatile algorithmic tool to enrich computer games. In this section, we present how procedural contents have previously been used games. We also examine how the potentially abundant variety generated by PCG can be dealt with to meet individual user requirements.

3.1 Categorizing PCG in Games

In most cases, not the complete game, but only specific parts of it are created procedurally. While terrain (Génevaux et al. 2013; Togelius et al. 2011) and assets like vegetation are most common examples, almost all elements in a game can potentially be generated procedurally (Hosking 2013): characters, artificial intelligence or stories and quests. Even music can be generated procedurally. The game *Audiosurf* is a good example: It can potentially generate as many levels as there are music tracks in the world, as every level is generated based on a piece of music. Tunes and rhythm are used to not only generate the outlines of a level, but also to place items according to the beats, creating a game with only a minimum of graphical effort.

How can this vast field of appliances be categorized and ordered? Hendrikx et al. (2013) proposed the following structure to differentiate between types and levels of procedurally generated content:

- Game Bits: e.g., textures, sound
- Game Space: the game world
- Game Systems: complex relations between game objects
- Game Scenarios: e.g., Story.

Using this differentiation, Hendrikx et al. looked at the application of PCG in 22 of the more influential video games of the last 35 years (Table 1).

We believe that this system is an approach well suited to describe PCG use in game. Especially, if it the content-oriented perspective is complemented by a

Table 1 Structural overview on the use of PCG in Games (Hendrikx et al. 2013)

Games	Release	Game bits	Game space	Game systems	Game scenarios
Borderlands	2009	X			
Diablo I	2000		X		
Diablo II	2008		X		X
Dwarf Fortress	2006		X	X	X
Elder Scrolls IV: Oblivion	2007	X			
Elder Scrolls V: Skyrim	2011				X
Elite	1984		X	X	X
EVE Online	2003	X	X		X
Façade	2005				X
FreeCiv and Civilization IV	2004		X		
Fuel	2009		X		
Gears of War 2	2008	X			
Left4Dead	2008				X
.kkrieger	2004	X			
Minecraft	2009		X	X	
Noctis	2002		X		
RoboBlitz	2006	X			
Realm of the Mad God	2010	X			
Rogue	1980		X		X
Spelunky	2008	X	X		
Torchlight	2009		X		
X-Com: UFO Defense	1994		X		

Copyright granted by Mark Hendrikx

Table 2 Categorization of technical complexity of PCG use (Hendrikx et al. 2013)

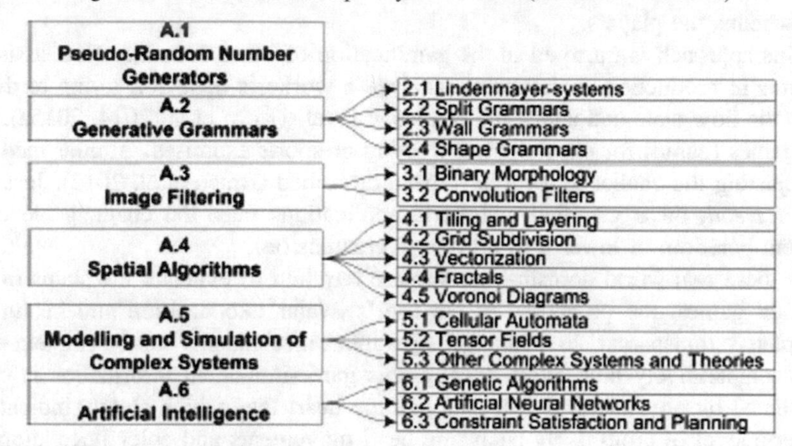

Copyright granted by Mark Hendrikx

technical perspective. Indeed, in a second step, Hendrikx et al. categorize PCG use according to different levels of technical complexity, beginning with pseudo-random number generators up to artificial intelligence (Table 2).

It would be interesting to investigate which technical PCG approaches were used in each game, e.g., in the games presented in Table 1. However, implementation details on games often remain unpublished, as the scientific documentation of algorithms and methods plays a minor role in the game industry. This situation is starting to change, as several of the other chapters in this book show.

3.2 *Adjusting the Challenge When Using PCG*

In the previous subsection, we showed a way to structure the potential abundance of procedural content. However, how can game designers prevent that this abundance overwhelms the players? How can the potentially endless variety of PCG be controlled and adjusted to the rest of the game world and the player requirements? In the field of artificial intelligence, Valve has drawn some attention with the series *Left 4 Dead*. Using the tool *AI Director*, they changed the dynamic creation of the game world depending on the course of the game. This was done in a covert way, e.g., by enabling passages that were previously blocked. In contrast to many competitors, they not only changed the behavior of the enemies, but also the game map according to the flow of the game, a method described by Thue and Bulitko (2012). Additionally, according to the calculated stress level of the player, enemy waves were adapted either to challenge the player, or to give him or her time to relax. This context-aware adaptation can prevent that the potential richness of PCG overwhelms the players.

This approach is mirrored in the gamification of other domains: New assistive systems in production environments analyze a worker's behavior, trying to determine the flow state and adjust the challenge level (Korn et al. 2014, 2015a). For exergames (games for physiological training or sports exercises), similar methods of adjusting the challenge level have been described (Brach et al. 2012). Just like *Left 4 Dead*, these gamified real-world applications raise the challenge level to prevent boredom or lower it to prevent overextension.

In these real-world domains, sensors are required to evaluate the status of the user. In games, the properties of the user's avatar like position and health are completely transparent, so challenge adjustment mechanisms can be implemented with comparatively little effort. Ideally, this game-immanent information is complemented by physiological user data like the heart rate, which clearly indicates a player's level of arousal. By analyzing head movements and color fluctuations in the face, pulse can even be measured without physiological sensors (Balakrishnan et al. 2013)—which leads us to the future of PCG.

4 The Future of PCG

While PCG can be applied in various content types, it also benefits from new technological platforms: For mobile games, it is especially rewarding to use procedural and dynamic approaches. Mobile or web-based games are often developed in small-scale productions (Korn et al. 2015b), so the benefits of using PCG are highest. Furthermore, many successful mobile games like *Doodle Jump* or *Tiny Wings* incorporate both a simple gameplay and a high replay value—features well suited for the use of PCG.

As the very successful game series *Borderlands* exemplifies, PCG can even be used in AAA productions to generate diversity in the gameplay. In *Borderlands,* a countless number of different weapons are generated procedurally. This creates the illusion of almost unlimited technical variety, enriching the game experience. *Minecraft* takes a step beyond the procedural generation of game elements and creates the complete game world procedurally. The resulting freedom to constantly discover new areas of the world in combination with the expensive ways to manipulate the game world generates a very high replay value (Belinkie 2010). Thus, the procedural approach is corresponding well with a central paradigm of game development: "the game enables the experience, but it is not the experience" (Schell 2015).

In spite of these two highly successful examples, there still are only a few studios relying on PCG, especially with bigger productions. When evaluating the consequences of PCG use, productions with a budget of several millions dollars usually still demand a higher level of reliability that counteracts innovation, flexibility and the "openness" resulting from procedural approaches. We think that AAA productions involving PCG will only become more frequent if the experimental factor concerning the rules of the game remains manageable. *Borderlands* has already cleared this hurdle: It neither changed the ground rules of the shooter genre nor let the procedural elements significantly alter the course and the rules of the game. Alternatively, the context-aware adaption mechanisms described in the previous subsection show effective ways to control the potential abundance of PCG in larger projects.

Nevertheless, the reluctance of larger studios regarding PCG currently puts independent (Indie) developers in the leading role of innovation in the game industry. Several of these smaller studios have shown the potential to create fascinating game worlds using PCG. In our view, PCG has not yet arrived at the peak of its potential. However, its importance is growing as the fascination of players for "endless" worlds and high replay values grow. Thus, the future of PCG in game design is mainly forged by the way it is applied: as a mere tool to create a variety of assets—or as a gameplay paradigm with a deep impact on user experience and replay value.

5 Conclusion

In this chapter, we portrayed the historical and mathematical background of procedural algorithms like fractals, Mandelbrot-sets and Perlin noise. A structural overview of games applying PCG as well as types of PCG was presented. We discussed ways to adapt the challenge to prevent that the potential richness and complexity of procedurally generated assets overwhelm individual players. Finally, we took a brief look at the future of PCG and explained why currently rather Indie developers than established game studios shape this future.

References

Balakrishnan, G., Durand, F., & Guttag, J. (2013). Detecting Pulse from Head Motions in Video (pp. 3430–3437). IEEE. 10.1109/CVPR.2013.440

Belinkie, M. (2010, November 11). What Makes Minecraft So Addictive? Retrieved from http://www.overthinkingit.com/2010/11/11/minecraft-vs-second-life/

Boeing, G. (2016). Visual Analysis of Nonlinear Dynamical Systems: Chaos, Fractals, Self-Similarity and the Limits of Prediction. *Systems*, *4*(4), 37. 10.3390/systems4040037

Brach, M., Hauer, K., Rotter, L., Werres, C., Korn, O., Konrad, R., & Göbel, S. (2012). Modern principles of training in exergames for sedentary seniors: requirements and approaches for sport and exercise sciences. *International Journal of Computer Science in Sport*, *11*, 86–99.

Génevaux, J.-D., Galin, É., Guérin, E., Peytavie, A., & Beneš, B. (2013). Terrain Generation Using Procedural Models Based on Hydrology. *ACM Trans. Graph.*, *32*(4), 143:1–143:13. 10.1145/2461912.2461996

Hendrikx, M., Meijer, S., Van Der Velden, J., & Iosup, A. (2013). Procedural Content Generation for Games: A Survey. *ACM Trans. Multimedia Comput. Commun. Appl.*, *9*(1), 1:1–1:22. 10.1145/2422956.2422957

Hosking, C. (2013, October 12). Stop dwelling on graphics and embrace procedural generation. Retrieved from http://www.polygon.com/2013/12/10/5192058/opinion-stop-dwelling-on-graphics-and-embrace-procedural-generation

Jones, O. (2013, September 1). Polygons to Pixels: The Resurgence of Pixel Games. Retrieved from http://gamingillustrated.com/polygons-to-pixels-the-resurgence-of-pixel-games/

Korn, O., Funk, M., Abele, S., Hörz, T., & Schmidt, A. (2014). Context-aware Assistive Systems at the Workplace: Analyzing the Effects of Projection and Gamification. In *PETRA'14 Proceedings of the 7th International Conference on PErvasive Technologies Related to Assistive Environments* (p. 38:1–38:8). New York, NY, USA: ACM. 10.1145/2674396.2674406

Korn, O., Funk, M., & Schmidt, A. (2015). Assistive Systems for the Workplace: Towards Context-Aware Assistance. In L. B. Theng (Ed.), *Assistive Technologies for Physical and Cognitive Disabilities* (pp. 121–133). IGI Global. Retrieved from http://services.igi-global.com/resolvedoi/resolve.aspx?doi=10.4018/978-1-4666-7373-1

Korn, O., Rees, A., & Schulz, U. (2015). Small-Scale Cross Media Productions: A Case Study of a Documentary Game. In *Proceedings of the ACM International Conference on Interactive Experiences for TV and Online Video* (pp. 149–154). New York, NY, USA: ACM. 10.1145/2745197.2755516

Mandelbrot, B. B. (1982). *The fractal geometry of nature*. San Francisco: W.H. Freeman.

Olano, M., Hart, J. C., Heidrich, W., Mark, B., & Perlin, K. (2003, March 1). Real-time shading languages. Retrieved from http://www.csee.umbc.edu/~olano/s2002c36/ch02.pdf

Perlin, K. (1985). An Image Synthesizer. In *Proceedings of the 12th Annual Conference on Computer Graphics and Interactive Techniques* (pp. 287–296). New York, NY, USA: ACM. 10.1145/325334.325247

Perlin, K. (1999, September). *Making Noise*. Presented at the GDCHardCore.

Persson, M. (2011, March 19). Terrain generation, Part 1. Retrieved from http://notch.tumblr.com/post/3746989361/terrain-generation-part-1

Schell, J. (2015). *The Art of Game Design: A Book of Lenses* (Second edition). Boca Raton: CRC Press.

Shaker, N., Togelius, J., & Nelson, M. J. (2016). Fractals, noise and agents with applications to landscapes. In *Procedural Content Generation in Games* (pp. 57–72). Springer International Publishing. 10.1007/978-3-319-42716-4_4

Thue, D., & Bulitko, V. (2012). Procedural game adaptation: Framing experience management as changing an mdp. In *Eighth Artificial Intelligence and Interactive Digital Entertainment Conference*. Retrieved from http://musicweb.ucsd.edu/~sdubnov/Mu270d/AIIDE12/03/WS12-14-012.pdf

Togelius, J., Kastbjerg, E., Schedl, D., & Yannakakis, G. N. (2011). What is Procedural Content Generation?: Mario on the Borderline. In *Proceedings of the 2nd International Workshop on Procedural Content Generation in Games* (p. 3:1–3:6). New York, NY, USA: ACM. 10.1145/2000919.2000922

Yannakakis, G. N., & Togelius, J. (2011). Experience-Driven Procedural Content Generation. *IEEE Transactions on Affective Computing*, *2*(3), 147–161. 10.1109/T-AFFC.2011.6

Ziegler, G. M. (2013). *Mathematik - Das ist doch keine Kunst*. Munich, Germany: Knaus.

Author Biographies

Michael Blatz is Chief Executive Officer (CEO) of KORION GmbH, a software company focusing on games and simulations based in southern Germany.

Blatz studied Virtual Reality and Game Development at the SRH Heidelberg University. He developed games for several platforms, especially mobile and PC. In 2013, he joined the Fraunhofer spin-off KORION. He participated in two national research projects and gained experience in several industrial game development projects. In 2016, he became CEO of KORION.

Blatz is a game enthusiast and an excellent connoisseur of game development strategies and methods on all platforms. His passion for platformers led him to procedural algorithms, e.g. for dynamic level generation. He applies procedural methods in most of KORION's projects—most recently in the Science Fiction strategy game Space Battle Core, where players explore procedurally generated solar systems.

Oliver Korn is a full professor for Human–Computer Interaction (HCI) at Offenburg University in Germany. He also is a certified project manager (German Chamber of Commerce, DIHK), professional member of the ACM and the IEEE, and an evaluator for the European Commission.

After completing his master's focusing on computational linguistics, he worked at the Fraunhofer Institute for Industrial Engineering (IAO) and the Stuttgart Media University (HdM). He focused on HCI, especially simulations, gaming, and gamification. In 2003, he co-founded KORION, a Fraunhofer spin-off developing simulations and games. As CEO, he was in charge of several national research projects and gained experience in countless industrial software development projects, both in the area of entertainment and business intelligence. In 2014, he received his Ph.D. in computer science at the SimTech Excellence Cluster of the University of Stuttgart. His research is published in numerous international publications.

He has been a game enthusiast since the times of the ZX81 and loves games with dynamic and procedural contents like *Civilization* or *Master of Magic*. Today, he works on the convergence of digital technology and real life, focusing on affective computing, gamification, augmented work, and learning. Korn's overall vision is integrating gameful design in education, health and work processes, augmenting and enriching everyday activities. In the area of affective computing, he aims to assess emotional states to improve computational context awareness.

Procedural Content Generation in the Game Industry

Alba Amato

Abstract Game content construction and generation are laborious and expensive. Procedural content generation (PCG) aims at generating game content automatically using algorithms, reducing the cost of game design and development. PCG systems have the potential to act as "on-demand game designers," but need to be as flexible as possible while creating content that meets the needs of designers and players. The aim of the chapter is to (1) analyze critically the utilization of PCG in the Game Industry; (2) present a taxonomy for PCG games; (3) address problems in applying PCG and discuss aspects of various solutions.

1 Introduction

Game content construction and generation are laborious and expensive. Procedural content generation (PCG) aims at generating game content automatically using algorithms, thus reducing the cost of game design and development (Risi et al. 2014). Moreover, PCG can also provide a way to generate personalized games that can adapt content according to a player's preference and optimizing their gaming experience. For those reasons, PCG is getting increasingly popular in game development field (Yannakakis and Togelius 2015).

The overall aim of this chapter is to analyze critically the utilization of PCG in the game industry, addressing problems and discussing the important aspects of the different solutions. However, I will not cover all those aspects. Yannakakis and Togelius (2015) arrived at a list of eight research challenges for procedural content generation:

A. Amato (✉)
Second University of Naples, Caserta, Italy
e-mail: alba.amato@unina2.it

© Springer International Publishing AG 2017

O. Korn and N. Lee (eds.), *Game Dynamics*,
DOI 10.1007/978-3-319-53088-8_2

- to create content generators that can generate content that is purposeful, coherent, original and creative
- to create a content generator that can create content in a particular style that it has somehow learned or inferred
- to generate multiple types of content for multiple games as almost all existing PCG algorithms generate a single type of content for a single game, the reusability of developed PCG systems is very limited and there is no plug-and-play content generation available
- to create tools to aid in adapting a particular representation for a particular class of content so that a particular representation could be adapted to a particular content class more easily
- to devise workable methods for communication and collaboration between algorithms working on generating different aspects or layers of the same artifact;
- to create believable animations even if the underlying characters are not procedurally generated;
- to establish game/music systems that interact in a much more complex way than what we are used to see in most games;
- to define theory and taxonomy of PCG in order to explain the relative advantages of different approaches, why some content generation problems are harder than others, and which approaches are likely to work on what problems.

In this chapter, I will mainly focus on the last challenge as there are diverse approaches for generating content for games, which are used in different genres and, due to this wide variety, it is difficult to compare the approaches without a standard taxonomy.

In Sect. 2, I will present technical background of PCG. In Sect. 3, a taxonomy of approaches will be presented, as well as an attempt to provide a common terminology and classification. Section 4 will focus on the history of the utilization of PCG in game industry, overviewing some of the most relevant and common algorithms and techniques used to develop PCG systems, highlighting the advantages of the different solutions. The final part of this chapter will be dedicated to questions and challenge relevant for the future of PCG in the game industry.

2 What Is PCG and Why It Is Used in the Game Industry

Procedural content generation refers to the creation of game content automatically (or semi-automatically) through algorithmic means (Yannakakis and Togelius 2015). It is a methodology for automatic generation of content of an entity, typically a game using algorithms or processes which can produce, due to their random nature, a very wide range of possible content related to the considered entity. The keystone of this methodology is the concept of randomness: Using a few parameters, the application of PCG ensures the creation of a high number of possible contents of a game, all differing from each other. It is important to define the meaning of the word "content" because according to its definition we can distinguish whether a content falls within the domain of the PCG or in other domains. The results from the application of PCG algorithms can be all kinds of elements affecting the gameplay: terrain, maps, layers, stories, dialogues, quests, characters, rules, dynamics, or weapons.

There is still not a clear distinction between PCG and PG (procedural generation), and often the two terms are used as synonyms, since they are based on the same algorithms to achieve different results. The main difference lies in the results generated: PCG generates content consisting of several components, while each individual component can be generated from the PG. PG also creates content that does not affect the gameplay directly, for example, procedural textures and procedural animation. A texture created by procedural generation techniques relates to PG, while the set of textures that form the setting of a level created by procedural generation techniques relates to PCG.

PCG has the potential to radically change how we conceptualize games, but also faces significant challenges regarding its integration into design practice. In recent years, it has become a mainstay of game AI, and significant research is directed toward the investigation of new PCG systems, algorithms and techniques (Khaled et al. 2013). In fact, there are a number of reasons for using and developing PCG techniques in games. The production of a game with high-level content implies a big effort in terms of cost and time. Normally one would expect that, with the growth of technology and the introduction of ever faster and more functional applications, this process of creation had at least speeded up, or even at least partially been automated. However, while technology in videogames is highly advanced, the creation of realistic 3D environments, rich dynamics and details of high-level graphics and content's creation in general are still largely manual. Usually, a team of people from different production departments of the game (programmers, sound engineers, artists, etc.) is responsible for the manual creation of all game contents. This has a significant effect on the budget available for the development, especially when there is the necessity to create a high quality game; as a result, content creation is viewed as the "bottleneck" for the total budget of a game.

The bottleneck created by the high costs and long development times of the game represents a barrier to the artistic and technological progress of games on any platform. Another reason for increasing PCG utilization is that it can potentially create infinite games. This happens when it is used for the generation of content in real time with a sufficiently high degree of variety. The history of games shows that this mechanism can lead to success since the early 1980s, when the *Rogue* videogame opened the way for PCG. *Rogue* is generally credited with being the first "graphical" adventure game. Its biggest contribution, and one that still stands out to this day, is that the adventure in *Rogue* was generated algorithmically. Every time you played, a new adventure was created. That is really, what made it so popular for several years in the early eighties. The vision of creating endless games attracted many developers, and this method was imitated numerous times over the years, for example, in the hit game series *Diablo*.

The use of mass storage is another point in favor of the use of procedural content generation. This problem was most significant in the early 1980s, when memory limitations of the existing end-user platforms did not allow the distribution of large amounts of predesigned content such as game levels. This problem reappeared 20 years later with the first mobile games. The advantage of using PCG is that content represented procedurally is "compressed" as long as it is not required by the gameplay. The adventure game *Elite* is the example par excellence: It could handle thousands of star systems in a dozen kilobytes of memory, representing only a few numbers of planets in compressed form. In expanded form, each planet had a name, population, prices of raw materials and so on. Even today, the use of storage should always be taken into account: Despite the large increases in modern hard drive capacities and the decrease in the price per unit, just a few video games with many GB each force many users to uninstall some games or to buy new storage devices to make space for new ones.

An example that addresses this memory problem is the game. *.kkrieger 1*, a 3D first person shooter, similar in kind to *Halo 3*. Unlike the latter, it uses procedural generation techniques to create textures, meshes and sounds which then are combined to form complete setting. *.kkrieger* makes extensive use of procedural generation methods: Textures are stored via their creation history instead of a per-pixel basis, thus only requiring the history data and the generator code to be compiled into the executable, producing a relatively small file size. Meshes are created from basic solids such as boxes and cylinders, which are then deformed to achieve the desired shape—essentially a special way of box modeling. These generation processes account for the extensive loading time of the game, as all assets relevant for the gameplay are produced during the loading phase. The entire game uses only 97,280 bytes of disk space—about 3 to 4 orders of magnitude less than in a similar game like *Halo 3*. According to the developers, *.kkrieger* itself would take up around 200 GB of storage space if it was stored in a conventional way.

Finally, PCG can increase the limits of human imagination. The algorithms may create new rules, levels, stories, which, in turn, can inspire designers to create new games (Fig. 1).

3 A Taxonomy of PCG

PCG has probably been used by too many people with too many different perspectives to arrive at a definition of procedural content generation that everybody agrees on. A graphics researcher, a designer in the game industry and an academic working on artificial intelligence techniques would be unlikely to agree even on what "content" is, and even less on which generation techniques are interesting. We can group, according to the latest classification of Hendrikx et al. (2013), the type of content that can be produced by procedural generation techniques into five main classes:

Fig. 1 Screen-shot of.kkrieger

- **Game Bits** are elementary units of the game content that do not affect the player's gameplay when considered independently. Included in this category: textures, sounds, vegetation, structures, behaviors, fire, water, stone, or clouds.
- **Game Space** represents the environment in which to play. It consists of different units Game Bits. Included in this category: indoor maps, outdoor maps, water bodies such as seas, lakes, or rivers.
- **Game System**, for example, ecosystems, road networks, urban planning.
- **Game Scenarios**, in which events occur, e.g., puzzle, storyboard, the history, the concept of levels.
- **Game Design**, including rules and objectives.

The type of algorithm suited for the creation of an element of a class depends on the type of element. Different algorithms can be used, starting with a simple pseudorandom number generator, to generative grammars up to the most advanced evolutionary programming techniques and artificial neural networks.

In Togelius et al. (2011a, b) another taxonomy for procedural content generation is proposed, centering on what kind of content is generated, how the content is represented and how the quality/fitness of the content is evaluated; search-based procedural content generation in particular is situated within this taxonomy. This work also contains a survey of papers in which game content is generated through search or optimization and ends with an overview of important open research problems. Shaker et al. (2015) present a revision of the above taxonomy that currently represents an important point of reference for many researchers.

- **Online—Offline Generation**: The first distinction concerns whether the content is generated online (while the game is running) or offline (during development of the game). An example of online PCG is when the player opens a door of a structure and the game instantly generates the interior: rooms, walls, decorations, etc. An example of offline PCG is when an algorithm creates the basic internal layout of a structure, which is then modified and refined by a designer before the game is completed. Clearly, the online PCG must comply with the basic requirements for creating valuable content: It must be fast, and the content must be qualitatively acceptable (depending on context).

- **Necessary—Optional Content**: The second distinction refers to the importance of the content, if it is necessary or optional for the particular game. Necessary content is required by players to progress through the game. For instance, dungeons that have to be crossed, monsters that must be defeated, crucial rules of the game, etc., fall into this category. Optional content is everything that the player can choose to not consider, such as weapons or facilities that can be ignored. Another difference between these two variants is the required quality: Necessary content must always be of excellent quality and functionally correct. It is not acceptable to create an unpassable dungeon, incorrect rules or unbeatable monsters if these anomalies make it impossible to continue the game. It is also not acceptable to generate the content that does not have a fair difficulty compared to the rest of the game. On the other hand, for optional content, it is principally allowed that an algorithm also produces unusable weapons or an unreasonable dungeon layout if the player can choose to discard the weapon or leave the dungeon.
  Clearly, the importance given to a content is dependent on the design of the game and its storyline. For example, the first person shooter *Borderlands* has a random generation mechanism of weapons, many of which are not useful, but analyzing these weapons is part of the heart of the gameplay. On the other side, a structure with a simple and low quality level of detail seems artificial and can make no credible setting in a game that as a main objective has visual realism, such as *Call of Duty 4: Modern Warfare*. It is interesting to note that some content may be optional in a category of games and necessary in others (for example, dungeons). Thus, the analysis of what is "optional" in a game needs to be made on a case-by-case basis.
- **Control Degrees**: This distinction depends on the type of generation algorithm that is used and how it can be parameterized. At one extreme, the algorithm can simply use a randomly generated number as an input; at the other extreme, the algorithm can use as a multi-dimensional vector containing the parameters that specify the properties of the content to be generated. In the case of the creation of a dungeon, the algorithm can use as input parameters the number of rooms, corridors branching factor, treasures, etc. The more degrees of control there are, the more the generated content can be customized and controlled.
- **Generation Deterministic or Stochastic**: This distinction relates to the degree of randomness in the build process. It is possible to conceive deterministic algorithms that either generate the same content with the same input parameters, or different content even with identical input parameters, as, for example, the *Rogue* dungeon generation algorithm.

Table 1 Revised taxonomy as presented by Shaker et al. (2015)

Online–offline
Necessary content—optional
Control degrees
Generation deterministic or stochastic
Constructive or generative algorithms-with-test

- **Constructive or Generative-with-test**: This last distinction refers to the output of the algorithm. Constructive algorithms generate the content and end their execution, producing the output result. For example, the algorithm of the Markovian process is typically a constructive PCG algorithm that produces stochastic content. However, it is necessary that there is a form of control in such a way as to avoid the production of unsuitable content. This can be done by operations or sequences of operations, which ensure that it cannot produce non-usable material. An example is the use of fractals to generate terrains. A "generative-with-test algorithm" incorporates a mechanism of generation and a test. This means that during the execution of the algorithm, each content instance is tested according to some criteria (which depend on the design of the game). If the candidate fails the test, it is fully or partially discarded and regenerated, and the process continues until the content does have sufficient quality (Table 1).

4 History of PCG in Game Industry

As described in Sect. 2, PCG was born as a way to compress data. *Akalabeth: World of Doom*, designed by Richard Garriott, was the first game to use a seed to generate the game world. It is a roleplaying game published by California Pacific Computer Company for the Apple II in 1980. The famous game *Rogue* (1980) is an open-source, free-to-change game that spawned popular free-to-use games such as *Moria* (1983), *NetHack* (1987) or *Angband* (1990). It is considered the ancestor of the Rogue-like games, which are one of the main game genres to use PCG in modern videogame to create an "endless" game experience (Fig. 2).

The Sentinel (1985) used PCG to create 10,000 unique levels stored in only 48 and 64 kilobytes. A procedural generation algorithm is used which generates each landscape from a small data packet: The 8-digit number given at the completion of a previous landscape. The number of landscapes was quite arbitrary, given the generation algorithm and was chosen as a balance between giving the players good value while not overwhelming them with an unreachable goal. Not all the landscapes were actually tested, but it was always possible to skip a difficult level by completing an earlier one with a different amount of energy.

Elite (1985) used PCG to generate a universe with 8 galaxies and 256 solar system each. Each solar system has 1–12 planets, each with a space station in its

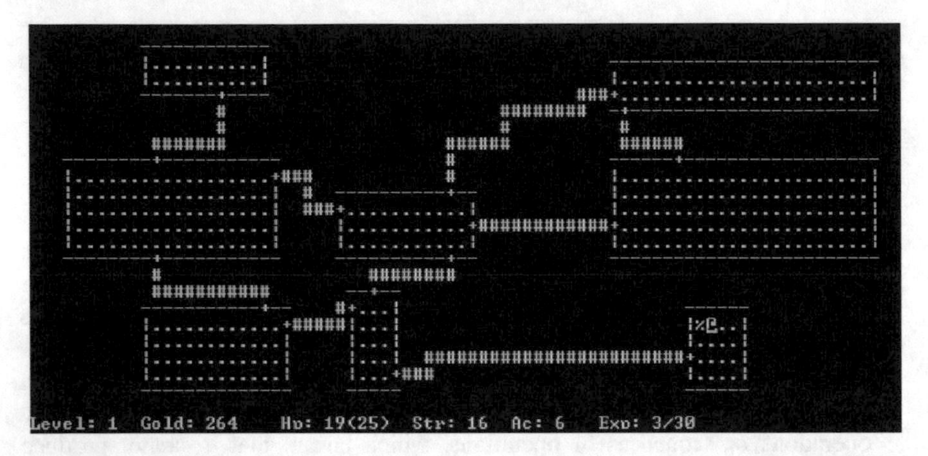

Fig. 2 Screenshot of Rogue

orbit, a name, a terrain, prices of commodities and local details. Due to the limited capabilities of 8-bit computers, these worlds are procedurally generated. A single seed number is run through a fixed algorithm the appropriate number of times and thus creates a sequence of numbers determining each planet's complete composition (position in the galaxy, prices of commodities, name, etc.). Text strings are chosen numerically from a lookup table and assembled to produce unique descriptions, such as a planet with "carnivorous arts graduates". This means that no extra memory is required to store the characteristics of each planet, yet each one is unique and has fixed properties. However, there also were some problems. Some solar systems were poorly connected, and the random name generator sometimes used profanity to name planets or space stations. The new version of *Elite*, *Elite Dangerous* uses PCG to generate a 1:1 replica of the Milky Way with more than 400 billion star systems. If we assume a storage requirement of just 1 KB per star system (very optimistic), without PCG, the full system would occupy more than 400 Terabyte (Aversa 2016).

With the establishment of CD-ROMs in 1980s, developers became able to store more data, so using procedural generation to build large worlds became less essential. During the 1990s, PCG was utilized to automatize designing assets (e.g., trees, rocks, foliage) to increase the amount of contents in a game in order to provide an improved replay value (Aversa 2016). *Diablo* (1996), shown in Fig. 3, was developed 16 years after the release of *Rogue* by Blizzard Entertainment and was one of the first games to bring the Rogue-like genre into the modern age. *Diablo* did a lot of things right, but as far as procedural generation goes, it popularized two specific gameplay elements: random dungeon layouts and random item or "loot" generation. Just like *Rogue* and all of its clones, *Diablo* generated dungeons according to an algorithm with random elements, but it took the process a step further: Instead of simply matching ASCII characters, *Diablo's* dungeons were created using 2D isometric graphics. By applying elements of randomization to its

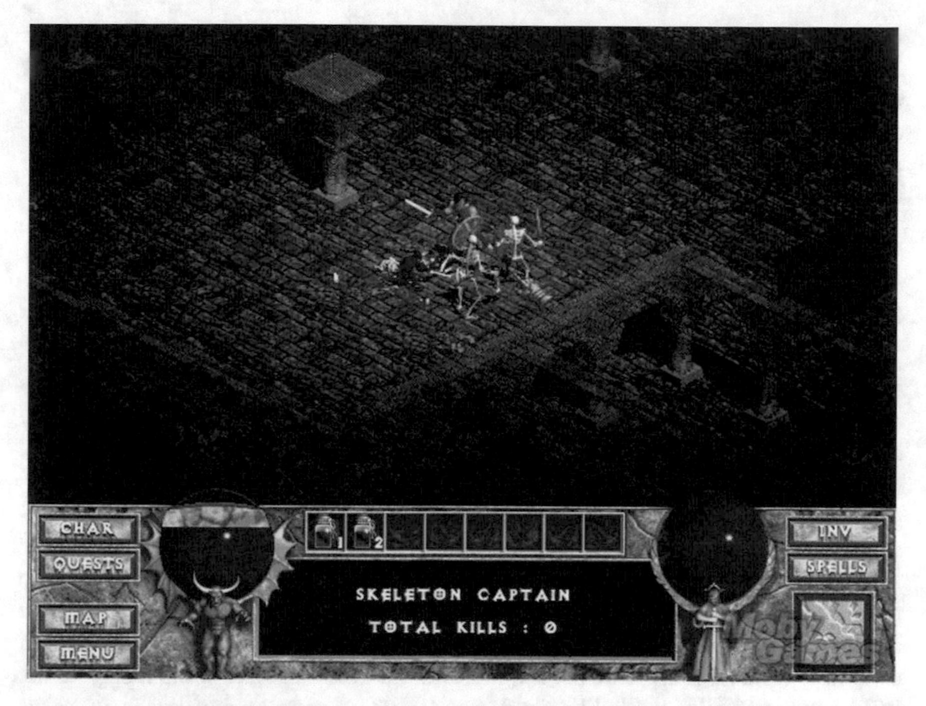

Fig. 3 Screenshot of Diablo (1996)

item system, it pioneered a type of gameplay still common today, e.g., in *Borderlands* or *Torchlight*. Color-coded tiers of rarity categorized items and each item's stats were generated on-the-fly (Lee 2016): a novelty in commercial videogames.

After 1996, commercial design tools using PCG were established, so PCG was no more confined to roleplaying games (RPG), space games or Rogue-like games. This gave rise to the modern age of PCG. The main idea behind procedural content generation is that game content is not generated manually by human designers, but by computers executing a well-defined procedure (Hendrikx et al. 2013). However, today PCG techniques are used to design level contents, helping designers to create game contents more quickly—and often without being noticed by the player as a design technique. When used more intensely, procedural techniques are an alternative to create game worlds in a limited amount of time without putting a large burden on the content designers. Such an intense use of PCG techniques is the generation of game levels/world while the game is being played or loaded: In this case, PCG instantiates the game objects, such as trees, monsters, characters, items, treasures and so on. Such techniques are used in games like *Elite*, *Minecraft*, *Spelunky*, *Diablo*, and many more (Togelius et al. 2013).

In mods, editors or middlewares, users can manually change PCG parameters to generate personalized contents, share these parameters, and so on. Today several middlewares for PCG exist, for instance, *CityEngine* (2016) for the generation of urban environments or SpeedTree (2016) for the generation of detailed forests. *SpeedTree* is widely used in AAA games, for instance, *The Elder Scrolls IV: Oblivion* by Bethesda or *The Witcher 2*. Finally, PCG techniques are also applied to agent behaviors (Aversa 2016) in order to model dynamic systems such as weather, and group and crowd behavior. For example, S.T.A.L.K.E.R.: The Shadow of Chernobyl, contains one thousand non-scripted characters.

5 Conclusion and Challenges

In this chapter, we presented the most common ways in which PCG is used. We discussed its applications: from early games that used it to deal with strict memory constraints, to recent games that use it to reduce the costs of creating large amounts of content. We presented a literature survey of selected future challenges in PCG, concentrating on themes mentioned in multiple sources where some level of consensus seems to be reached.

Some established game studios are starting to use PCG instead of artists, to produce games faster and cheaper while preserving quality (Shaker et al. 2015). At the same time, for many indie studios, PCG is the only way to produce enough content to create a game with an adequate amount of content and some replay value. PCG can potentially work like on-demand game designers, but needs to be as flexible as possible, rapidly creating content that meets the needs of designers and players while reviewing the quality of that content.

In future, PCG will be highly used for game elements on mobile devices (smartphones and tablets). However, on all platforms automatic techniques for generating contents with an optimal level of difficulty are required to keep the player in a state of flow. While PCG is very strong when there is a dependence on cost and time factors, without appropriate constraints and controls it is unsure if PCG creates the expected play experience. This disadvantage keeps many producers of video games from using PCG. In future research, an important challenge is analyzing the data collected during play sessions in order to detect and correct defects in gameplay and improve the applicability of PCG. It is possible to hypothesize alternative environments that evolve by genetic algorithms, resulting in higher quality variants.

References

Risi, S., Lehman, J., DAmbrosio, D., Stanley, K. (2014). Automatically Categorizing Procedurally Generated Content for Collecting Games. In: *Proceedings of the Workshop on Procedural Content Generation in Games (PCG) at the 9th International Conference on the Foundations of Digital Games (FDG-2014)*

Yannakakis, G.N., Togelius, J. (2015). Experience-driven procedural content generation (extended abstract). In: *ACII, IEEE Computer Society 519–525*

Khaled, R., Nelson, M.J., Barr, P. (2013). Design metaphors for procedural content generation in games. In: *Proceedings of the SIGCHI Conference on Human Factors in Computing Systems. CHI '13*, New York, NY, USA, ACM 1509–1518
Procedural content generation (2016) Retrieved from: https://pcg.wikidot.com/
Togelius, J., Kastbjerg, E., Schedl, D., Yannakakis, G.N. (2011) What is procedural content generation?: Mario on the borderline. In: *Proceedings of the 2nd International Workshop on Procedural Content Generation in Games. PCGames '11*, New York, NY, USA, ACM (2011) 3:1–3:6
Hendrikx, M., Meijer, S., Van Der Velden, J., Iosup, A. (2013). Procedural content generation for games: A survey. *ACM Trans. Multimedia Comput. Commun. Appl. 9* 1:1–1:22
Togelius, J., Yannakakis, G.N., Stanley, K.O., Browne, C. (2011). Search-based procedural content generation: A taxonomy and survey. *IEEE Transactions on Computational Intelligence and AI in Games 3*, 172–186
Shaker, N., Togelius, J., Nelson, M.J. (2015). Procedural Content Generation in Games: A Textbook and an Overview of Current Research. Springer
Elite (2016). Retrieved from: https://en.wikipedia.org/wiki/Elite_(video_game)
Aversa, D. (2016). Procedural contents generation history and techniques used in the modern video-game industry. Retrieved from: www.davideaversa.it/tag/pcg/
Lee, J. (2016). How procedural generation took over the gaming industry. Retrieved from: http://www.makeuseof.com/tag/procedural-generation-took-gaming-industry/
Cityengine (2016). Retrieved from: https://en.wikipedia.org/wiki/CityEngine
Speedtree (2016). Retrieved from: https://en.wikipedia.org/wiki/SpeedTree
Togelius, J., Champandard, A.J., Lanzi, P.L., Mateas, M., Paiva, A., Preuss, M., Stanley, K.O. (2013). Procedural content generation: Goals, challenges and actionable steps. In: *Artificial and Computational Intelligence in Games*. 61–75

Author Biography

Alba Amato is currently a postdoc researcher in computer science at the Second University of Naples. She received her M.Sc. in computer science from the University of Napoli Federico II in 2007. She received her postgraduate degree from the School of Specialization in Computer Science Teaching of the University of Naples Federico II and her Ph.D. from the Second University of Naples.

Her research is focused on multiagent systems, Internet of things, smart grid and cloud computing. She is also interested in big data, artificial intelligence, and the intersection between artificial intelligence and games—in particular, algorithms for the procedural generation of games, levels, narrative, environments, animation, characters, and other game contents together with their historical utilization.

She co-authored more than 50 journal articles, book chapters, and refereed conference papers.

She participated in many national and international conferences and workshops. She is reviewer for several IEEE, Inderscience, and Wiley journals in the areas of computer science and for national and international conferences and workshops.

Design, Dynamics, Experience (DDE): An Advancement of the MDA Framework for Game Design

Wolfgang Walk, Daniel Görlich and Mark Barrett

Abstract Although the **Mechanics, Dynamics and Aesthetics (MDA) framework** is probably the most widely accepted and practically employed approach to game design, that framework has recently been criticized for several weaknesses. Other frameworks have been proposed to overcome those limitations, but none has generated sufficient support to replace MDA. In this chapter, we improve the MDA framework, place it on new pillars and thus present the **Design, Dynamics, Experience (DDE) framework** for the design of computer and video games.

1 Introduction

The **MDA framework** was designed and introduced by Hunicke et al. (2004) to "clarify and strengthen the iterative processes of developers, scholars and researchers alike, making it easier for all parties to decompose, study and design a broad class of game designs and game artifacts." Since then, the MDA framework has become one of the fundamental approaches to game design, being well cited and accepted especially in academia. They defined the eponymous parts of the MDA framework (**Mechanics**, **Dynamics** and **Aesthetics**, see Fig. 1) as follows:

W. Walk (✉)
Walk Game Productions, Karl-Leopold-Str. 6, 76229 Karlsruhe, Germany
e-mail: wolfgang.walk@t-online.de

D. Görlich
SRH University Heidelberg, Ludwig-Guttmann-Str. 6, 69123 Heidelberg, Germany
e-mail: daniel.goerlich@srh.de

M. Barrett
Ditchwalk, 1830 Glendale Rd, Iowa City, IA 52245, USA
e-mail: mark@ditchwalk.com

© Springer International Publishing AG 2017

O. Korn and N. Lee (eds.), *Game Dynamics*,
DOI 10.1007/978-3-319-53088-8_3

The MDA framework formalizes the consumption of games by breaking them into their distinct components:

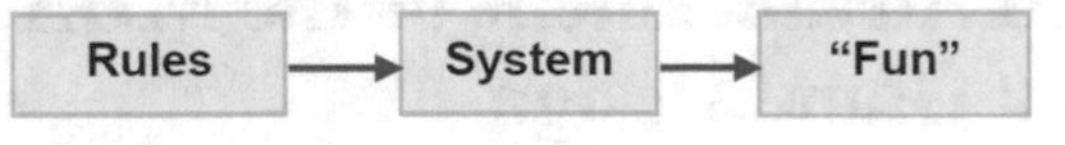

…and establishing their design counterparts:

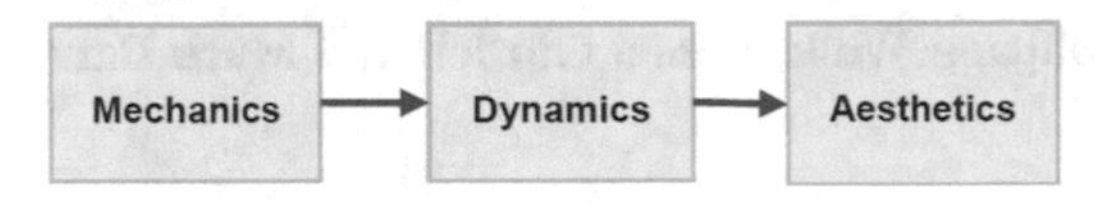

Fig. 1 The basic concept of the MDA framework by Hunicke et al. (2004)

- **Mechanics** describes the particular components of the game, at the level of data representation and algorithms.
- **Dynamics** describes the run-time behavior of the **Mechanics** acting on player inputs and each other's outputs over time.
- **Aesthetics** describes the desirable emotional responses evoked in the player, when she interacts with the game system.

Only recently, in 2015 and 2016, articles have been published in which MDA was deeply analyzed, criticized and challenged. MDA was first discussed in depth on Gamasutra by Silveira Duarte (2015). In that same year, an article by Polansky (2015) was followed by an article by Lantz (2015), both adding further points of criticism. In summary, these three authors identified two major weaknesses in the MDA framework:

1. It neglects many design aspects of games, focusing too much on game mechanics.
2. It is (therefore) not suitable for all types of games, including particularly gamified content or any type of experience-oriented design (as opposed to functionality-oriented design).

In addition, MDA fails to provide a framework or even a coherent approach for narrative design. Are narratives a component of **Mechanics** in MDA? In part, but not completely. Narrative design certainly does not belong to **Dynamics**—and also not to **Aesthetics** as defined by Hunicke et al. (2004). Inevitably, every attempt to apply MDA to narrative design stretches the framework to the breaking point.

However, MDA is not the only game design framework. Having identified weaknesses in MDA, several practitioners and scientists proposed their own approaches to game design. In this chapter, we present those counterproposals and their distinctive features, consolidating many of those improvements, along with our own developments, into a revisited MDA framework based on three new pillars: **Design**, **Dynamics and Experience (DDE)**.

2 State-of-the-Art Game Design Frameworks

One of the best-known foundations of game design was introduced by Jesse Schell. Called the **Elemental Tetrad** (Schell 2008), it has been gratefully accepted by academics as well as practitioners and is considered essential knowledge for every modern game designer. While both MDA and the Elemental Tetrad define **Mechanics** and **Aesthetics** as central components, they do not agree on how those components are connected.

In MDA, **Mechanics** describes the components of the game at the level of data representation and algorithms, **Dynamics** describes the run-time behavior of the **Mechanics,** and **Aesthetics** describes the desired emotional responses evoked in the player (see Fig. 1). Hunicke et al. conclude that MDA supports a formal, iterative approach to designing and tuning a game, allowing the game designer to reason explicitly about the goals of the **Aesthetics**, identify suitable **Dynamics** to support those goals, and then define the **Mechanics** accordingly. This way, the MDA framework introduces three levels of abstraction, implying that **Mechanics** and **Aesthetics** are only indirectly connected.

Schell's Elemental Tetrad does not contain this abstraction, implying the opposite, i.e., that **Mechanics** and **Aesthetics** are indeed directly connected. Furthermore, Schell introduces two new components, which MDA does not sufficiently address: **Story** and **Technology**. Although Schell makes clear that certain components of a game, including the story and the underlying technology used to realize that game, are less visible (or perceptible) to the player, he proposes that each of the tetrad's four components is directly connected to every other one (see Fig. 2).

Although the MDA framework and the Elemental Tetrad seem to contradict each other, Paul Ralph and Kafui Monu combined them into their MTDA+N working theory, employing **Mechanics**, **Technology**, **Dynamics**, **Aesthetics** and a **Narratives Framework** (Ralph and Monu 2014). However, this approach has not been widely adopted so far.

The MDA framework has also been criticized for giving the game designer only indirect control over the dynamics and aesthetics of a game, because it assumes that all of the game's dynamics and aesthetics result from its mechanics. What MDA does not account for, but also does not preclude, are additional aesthetical game elements, which are *not* produced by the game's mechanics or dynamics. Effectively, the MDA framework neglects the *purely* aesthetical requirements of a game or its players. Brian Winn therefore concludes that MDA "does not specifically address aspects of game design beyond the gameplay, including the storytelling, user experience, and influence of technology on the design" (Winn 2008). He also criticizes the MDA framework for focusing on games made primarily for entertainment, but does not consider the challenges specific to the design of serious games. Abstracting Winn's point of criticism further, we can conclude that the MDA framework is not suitable for all types of games.

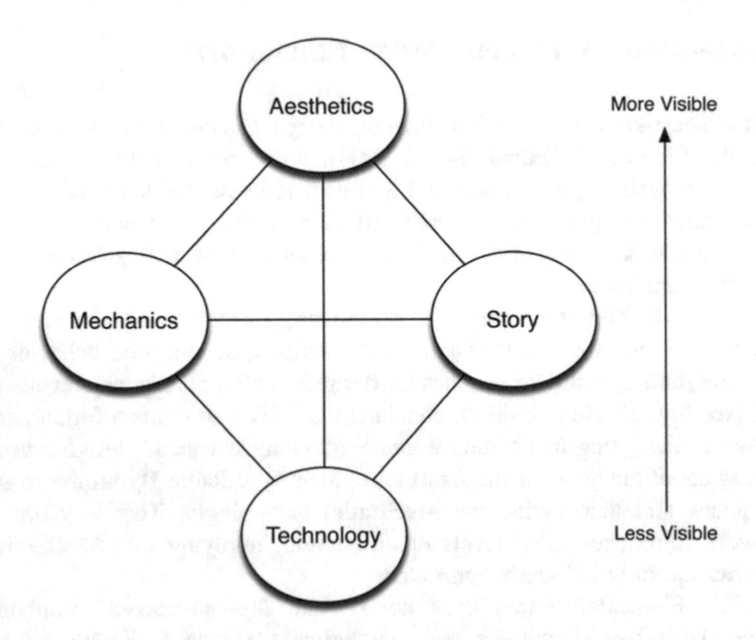

Fig. 2 The elemental tetrad by Schell (2008)

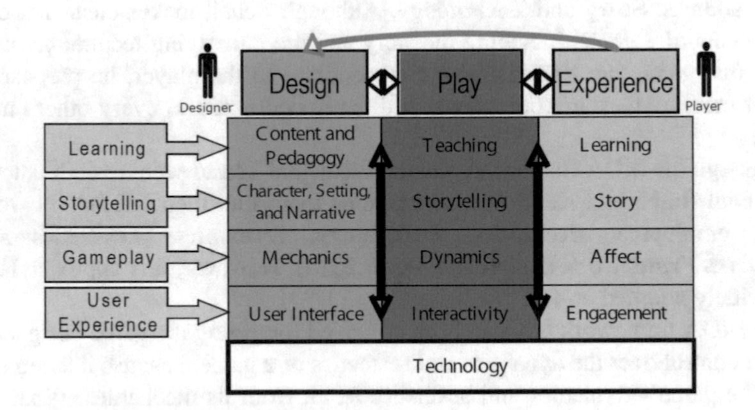

Fig. 3 The **Design**, **Play** and **Experience** (DPE) framework by Winn (2008)

Winn introduced a new concept by expanding MDA into the **Design**, **Play** and **Experience** (**DPE**) framework (see Fig. 3). DPE depicts the relationship between the game designer and the player in the same way as MDA, but also translates additional design aspects into layers. Designed mainly for what Winn terms "serious games," DPE proposes that such designs do not simply comprise gameplay mechanics, but also include pedagogical content to be learned; characters, settings and narratives of the story to be told; and a user interface. Finally, as per Schell, DPE also incorporates the underlying **Technology** as a fundamental prerequisite for realizing such a game and for mediating between the **Design**, **Play** and **Experience** aspects of a game. Similar to the proposal in this chapter, Winn replaces **Aesthetics** with **Experience**, acknowledging that the **Aesthetics** of a game are not directly received by the player, but experienced in an individual, subjective and unique way.

While the DPE framework proposes an iterative design process, especially for serious games, a significant number of frameworks already exist in the field of gamification:

- the Motivation Design process (Werbach and Hunter 2012),
- the Player-Centered Design methodology (Kumar and Herger 2013),
- the Octalysis gamification framework (Chou 2014),
- the Gamification Model Canvas (Jilménez 2014), which combines the Business Model Canvas (Osterwalder and Pigneur 2011) with the above-mentioned MDA framework, and
- the user-centered Gameful Design process (Herrmanny and Schmidt 2014).

3 Introducing the DDE Framework

The MDA framework is the most widely accepted framework for professional game design. To overcome its weaknesses, we propose the DDE framework, which rests on three new pillars of **Design**, **Dynamics** and **Experience**.

3.1 *From Mechanics to Design*

In MDA, **Mechanics** describes the particular components of a game at the level of data representation and algorithms. By analogy, this approach would describe an assembled car as "motor, gearbox and wheels"—yet cars are obviously meant to be, and marketed to be, much more. For example, for years BMW has advertised its cars with the slogan, "Sheer Driving Pleasure," because to the average customer a car is more than just the sum of its mechanical parts. Likewise, games are not simply the sum of their game mechanics. To a greater degree, most games are also aesthetical thanks to data representing (among other things) graphics and sound assets, which are specifically composed to create a world for the player to experience. Viewed from the gamer's perspective, games are (or should be) experiences rather than functional units.

The MDA terminology also conflicts with itself. The intent of **Aesthetics** was to describe "desirable emotional responses evoked in the player"—something completely different from data representation. In his article, Frank Lantz explained that in MDA terminology, **Mechanics** effectively means everything that is directly designed by the designer and under no direct influence by the player: "But even more problematic is the term 'mechanics.' Again, MDA wants to use this word broadly to refer to *all* of the stuff that the designer has control over–not just the rules of the game but the materials as well, the recipe *and* the ingredients." We agree with the interpretation of Lantz. To drive the point home that **Mechanics** does not really fit as a descriptor, here is an incomplete list of elements that designers have full control over, which accordingly belong to the MDA category of **Mechanics**:

- GAME CODE: Game Architecture, I/O, Game Objects, Game Rules (code level), Interface (code level), Interaction Design (code level), …
- GAME RULES (documentation): Structure, Balancing, Timing, Space, Plot Branching, …
- WORLD DESCRIPTION (documentation): World and Game Rules, Flora and Fauna, Societies, Characters, Religions, Laws, Physics, …
- STYLE (documentation): Graphics, Sound, Narrative, …
- FUNCTIONAL INTERFACE (data representation): Diegetic, Non-Diegetic, Spatial, Meta (Stonehouse 2014)[1]
- CONTENT INTERFACE (data representation): Interaction Design (interface level), Graphics and Sound, Narratives, …[2]

For a number of items on the above list, it is problematic to summarize them under the **Mechanics** label. Indeed, some of the items are not mechanical at all, such as the style of the graphics, which is an aspect of **Aesthetics**, yet that style is also under direct control of the designer before becoming part of the final game experience. If we further argue that a graphical style guide—i.e., the documentation which graphic designers rely on to perform their work—is not represented in the game itself on the level of data representation or algorithms, then the MDA model leaves no place for that crucial design element—again proving MDA to be incomplete as a framework.

We adopted two lenses for re-framing the MDA model. First, we focused on the production process of a game, especially with regard to its iterative nature. Our version of the model needed to allow for that. The second lens came from a seemingly different perspective which is in fact closely related: Why do game stories too often fall short? Our version of the model also accounts for narrative as an entity.

In rebuilding the MDA framework, our goal was to span the entirety of the development process, including both the actual production of a game and the

[1]According to Stonehouse (2014), "spatial" refers to user interface elements that may break the narrative, but are still a part of the game environment. "Meta" refers to user interface elements which do not fit into the geometry of the game environment and must therefore be displayed separately (usually in a 2D hub plane), yet still belong to the game's narrative. "Diegetic" refers to UI elements existing within the game world (fiction and geometry); "non-diegetic" is UI elements and events, which occur outside the game's fiction and geometry.

[2]In the DDE framework, everything that is displayed on screen or can be heard through the speakers is—along with any other feedback to the player—considered part of the game's interface, because it translates the abstract layer of code into something the player can understand. This has certain implications that will be addressed later in this chapter.

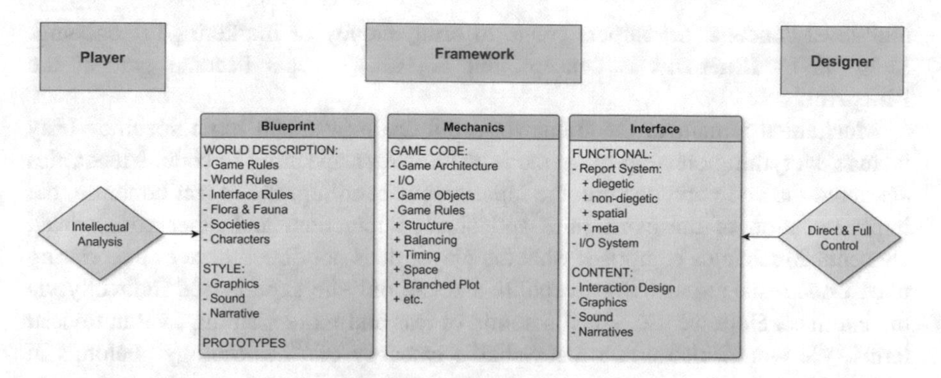

Fig. 4 The **Design** part of the DDE framework

player's eventual perceptive journey—neither of which the MDA framework addresses. Schell argued for the iterative production model in his Elemental Tetrad by pointing out the dependencies between his four design instances, but he failed to deliver a valid line of argument about why **Story** should be an instance of its own, intuitive as that may seem. In fact, one could easily find reasons to divide **Story** between **Aesthetics**, **Mechanics** and **Interface**, where the latter is not even part of Schell's model.

In deliberating a re-framed **Mechanics** category, we found that we needed three subcategories to make clear distinctions without blurry concepts lurking in between. Figure 4 shows our take on what Hunicke et al. called **Mechanics**, and what we call **Design**. The term "Design" was also proposed by Jesper Juul in the comments section following Frank Lantz's article—however, we make no claim that Juul would agree with our substructures. Every element in the figure is directly designed by designers who have full control over that stage of the development process. In reality, of course, having "full control" is often wishful thinking, since few designers have full control over every graphic, sound assert or line of code. Typically, they do not control all assets themselves but order specific changes that will directly affect those assets, and thus the design of the game.

3.2 Design Subcategories

Blueprint is manifested by that part of the **Design** dealing with the game world *in concept*: its cultures, religions, physics and other rule sets; the free form notation of the game mechanics; and the developed styles of art design, narrative design, character design and sound design that together create the aesthetical experience. In an earlier stage of our model, we used the term "Setting" for this substructure, but **Blueprint** seems much clearer as it reflects the dominance of planning and documentation throughout the development. Another term we evaluated was "Concept"—however, it was rejected because of potential confusion with the high-level concepts developers create to bring money or marketing/PR onboard. Thus, all of **Blueprint** is concept, but not all concepts become part of the **Blueprint**.

Mechanics remain in the framework, but are now much more specific. They include everything creating the game *in the abstract*, meaning in code. **Mechanics** are about the code architecture, the input/output handling, the object handling, the implementation of the game rules and object interaction, and other code-related elements. Mechanics comprises what the player does not directly see or hear during play. Code itself remains imperceptible and can only be experienced indirectly via the interface. Here we use Schell's notion of less and more visibility, yet in revised terms. We would also argue that Schell's category of "**Technology**" belongs in DDE's **Mechanics** subcategory as computer and video games are ultimately controlled by code on variant hardware.

Interface concerns the design and production of elements creating the game *in the concrete*: everything that serves to communicate the game world to the player—how it looks, how it sounds, how it reacts and interacts with the player and the game's internal feedback loops. **Interface** also contains the report system that every game needs, be it diegetic or non-diegetic, spatial or meta. Every graphic asset, sound asset, cut scene or text on display is part of the interface as long as it is also part of the game data. It is everything the player hears and sees, and every piece of data that does not belong to the executable or configurative code level of the game.

3.3 Dynamics

Building on the **Design** category, there were considerably fewer problems with the category **Dynamics**, though in practice we still need tools that grant more control over that part of the creative process, helping to minimize expensive design iterations. As Frank Lantz correctly noted: "For me, the greatest strength of MDA is that it emphasizes the 'second order' nature of game design. *Mechanics* is used to refer to the parts of the game that the designer has direct control over, *aesthetics* refers to the qualities of player experience that the game ultimately generates, and in between, linking the two, are the *dynamics* of the game in action—the behavior of the game's different parts interacting with each other and the player while the game is being played."

While there are significantly fewer problems to address, the DDE framework does add basic structures to MDA's definition of **Dynamics** in order to be more specific, especially with regard to the divergent perspectives of designers and players (Fig. 5).

Returning to the automotive metaphor, if **Design** involves planning all of the parts of a car, as well as assembling those parts, **Dynamics** defines what happens when the engine starts and all of those parts work together: the pistons, crank shaft and valves of the motor; the gear box; the suspension and springing of the seats—even the road, weather, driving style, tremor of the steering wheel, mechanical

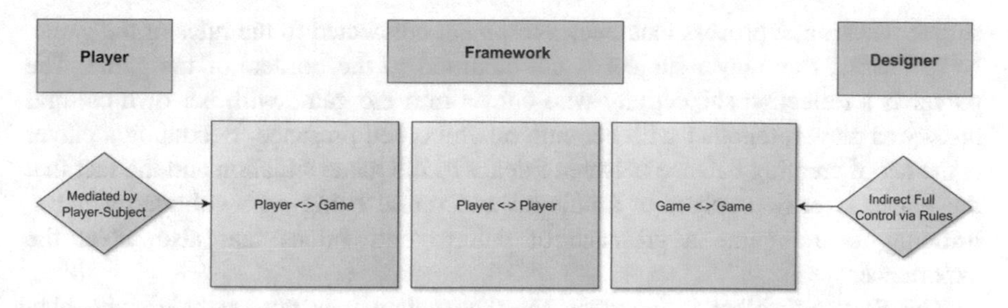

Fig. 5 The **Dynamics** part of the DDE framework

noises and the song on the car stereo are part of the dynamics. Obviously, such dynamics must be designed or at least be taken into consideration during the design process, meaning **Dynamics** as a part of the design framework is *still* under full control by the designer.

At least this is true in theory—assuming an iterative design process, which allows endless designing until all dynamics work as intended. In practice there will always be too much complexity and too many players to predict every behavior, but it might help designers in their decision making if they were aware of the difference between theory and practice at that stage of development. The most important difference between **Design** and **Dynamics** is that the designer's control over **Dynamics** is indirect, because every element the designer can directly affect falls into the category of **Design.** By definition, **Dynamics** always implies emergence and unpredictability.

Due to the interaction of the game system with itself, **Dynamics** as structured in our approach are by definition procedural. Depending on the particular design of a game, the procedural approach can result in gameplay which is more or less emergent. Later we will also discuss the cause and role of emergence relative to player perception.

3.4 *From Aesthetics to Experience*

The MDA framework describes **Aesthetics** as the desirable emotional responses which are evoked when the player interacts with the game system. While this seems like a clear definition, closer inspection reveals important instances and structures not mentioned by MDA, but processed by DDE, including the **Player-Subject** and the **Antagonist**.

While the *player* is confronted with the dynamics of a game, that confrontation is also indirect. As Miguel Sicart has shown, the indirect nature of interaction happens via the creation of the **Player-Subject** (Sicart 2009). In introducing the term, Sicart wrote: "I have already reasoned that games are processes. In the same line, it is possible to understand the act of playing a computer game as an act of subjectivization, a process that creates a subject connected to the rules of the game. Nevertheless, this player-subject is not confined to the borders of the game. The player is a reflective subjectivity who comes into the game with her own cultural history as player, together with her cultural embodied presence. Becoming a player is the act of creating balance between fidelity to the game situation and the fact that the player is only a subset of a cultural and moral being who voluntarily plays, bringing to the game a presence of culture and values that also affect the experience."

The **Player-Subject** is based on the theory that it is not really *us* who play games, but a subset of ourselves. One can imagine the **Player-Subject** as similar to a mental persona: a character existing solely in the mind, strongly influenced by us yet able to make decisions we might never make in real life. This mental character has a different set of abilities and a different set of ethics. As game players we often put our **Player-Subject** into mentally dangerous situations to experience that danger without exposing our real selves to harm. This allows experiencing and exploring ethically and mentally challenging situations from a safe place: We can rake in the benefits and rewards without risking the effects comparable real-life situations would produce. In the context of **Dynamics**, as well as the greater DDE framework, the designer and the **Design** do not deal with the player directly, but with the **Player-Subject**, because it is that instance or aspect of the player that makes decisions while playing.

Adopting the automotive metaphor, **Dynamics** is all of the parts of a car in unified action, plus any external influences, but it is *not* the driving experience. Inside the original MDA framework, the player's experiences would be referred to as **Aesthetics**, but there are a lot of reasons why **Aesthetics** is an even worse term than **Mechanics**—and why we consider **Experience** the superior term. We already pointed out that the MDA framework treats many aspects of deeply aesthetical design questions as **Mechanics**. In response, aesthetical decisions and data representations are now part of all three subcategories of our own category of **Design.** However, the problem with calling any category "aesthetics" in any such a framework runs deeper because "aesthetics" is also a philosophical term used in at least two distinct ways:

1. In phenomenology, trying to answer the question *how* we perceive things. Kant, for instance, claimed we cannot perceive anything independent from time and space, because these are a priori intuitions we cannot help but have.
2. Aesthetic theory, trying to answer the question *why* we perceive something as beautiful or unpleasant, as art or non-art. What is it that creates beauty in our mind?

Deepening the confusion, "aesthetics" is also a psychological term referring to *how* different people can and often do perceive the same color, sound, melody, picture or text in completely different ways, including trying to understand the reasons and implications behind those differences. What is it in our mind that determines the volatility of our perception? Finally, "aesthetical" is also an everyday synonym for "beautiful" or "well-shaped."

Clearly, even among (or particularly among) highly educated individuals, "aesthetics" as a term can confuse this side of the MDA design process. In fact, the term is controversial even among philosophers, and its use among artists was famously denounced—not without reason—by Barnett Newman: "*Aesthetics is for the artist as ornithology is for the birds*" (Crysler et al. 2012). Unfortunately, a framework loaded with philosophical terms that are used for assessing art *after* it has been created is of little developmental use. While ideally the DDE framework would support the critics as well, primarily it must help developers in their daily work. Such a practical framework should therefore focus on the goal of helping game designers, while also accurately representing the underlying art form.

Let us recall the definition in the MDA paper: "*Aesthetics describes the desirable emotional responses evoked in the player, when she interacts with the game system.*" Any reading of that definition involves time and space, in most cases hours and hours of play. Yet for the participants, something that happens over time and in space is necessarily a journey, an experience. By its very nature, game design is or should be experience design. This even is mentioned in the MDA paper: "*In addition, thinking about the player encourages experience-driven (as opposed to feature-driven) design.*" (Fig. 6).

That is why we, in line with Jesper Juul, suggest changing the MDA category from **Aesthetics** to **Experience**. This experience begins as soon as the player plays the game—or even earlier, when she first learns about the game, considers buying it, installs it, … Experience is also not separated in time from the dynamics, because dynamics need time and space as a priori *intuitions*. Only when game dynamics become evident can the **Player-Subject** experience them.

3.5 Antagonist and Player-Subject

Now we introduce another entity into our framework, which will help to understand the goal of game design. The **Antagonist** is well known from storytelling and other narrative art forms. In different guises, the **Antagonist** exists in every art form, be it the counterpoint in music or complementary colors in painting. The reason is obvious: It is through conflict, contrast or tension that almost all art generates interest at differing levels of awareness.

If a game design and its dynamics are sophisticated enough to create an **Antagonist** for the **Player-Subject**—i.e., the game appears to the **Player-Subject** as a monolithic entity presenting a challenge—then *the game itself* becomes a worthy opponent. As a result, playing the game over time creates a journey that works on multiple levels:

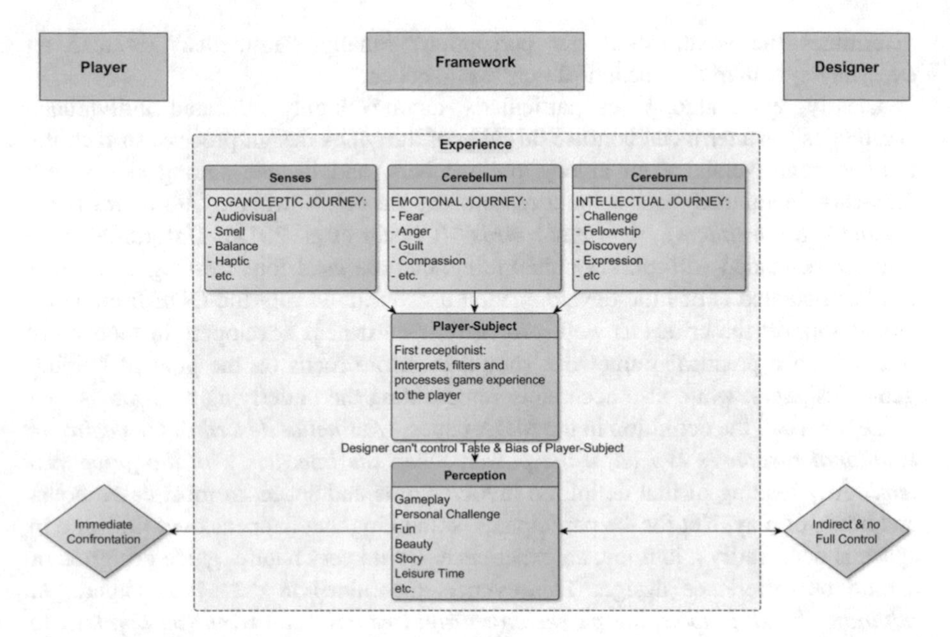

Fig. 6 The **Experience** part of the DDE framework

- **Senses**: The organoleptic journey consists of all the player's sensory experiences from start to finish. It is the totality of what the player sees, hears and senses through the output devices, and perhaps even from the surroundings.
- **Cerebellum**: The cerebellar journey consists of all the emotions the player experiences while playing the game: fears and horrors, sadness, guilt and anger, happiness, joy and any other emotions.
- **Cerebrum**: The cerebral journey consists of all the intellectual challenges and decisions the player experiences and consciously contemplates.

While we agree that the two latter subcategories do not fully reflect the physiological functionality of the eponymous parts of the human brain, they help to give the designer orientation. Together the three journeys and how they are processed by the **Player-Subject** are formative for the player's *perception* of the game. This perception in turn has many names and subconcepts, and of course, personal taste and bias play a large role in the individual game experience.

If we look at the role of the player, **Perception** is the level of immediate confrontation. However, the player's perception is based on the interpretation of the game world and subsequent decision making by the **Player-Subject**. This interpretation depends on the **Player-Subject**'s individual taste and current mood. The **Player-Subject** also works as a protective layer, enabling the player to dissociate herself from the game. Ultimately, the player's interpretation of the **Player-Subject**'s collective journey creates the immediate confrontation—the player's perception of the game.

From the perspective of the designer that is also the level where full control over the work is lost (even indirectly), because the individual perception of the player takes over. The player can sense, feel and think whatever she wants about the experience. She can like it or hate it, enjoy it or find it boring, get what she hoped for or did not expect, be challenged or taken to the emotional brink. The better the designers know the target audience, and the better the game targets that audience, the better it will be received. However, predicting whether a game will prove immersive still involves a degree of chance. This is also why building a game by objective design criteria cannot ensure its commercial success—although it helps rather than hinders—and why knowing the emotional expectations of the audience is critical to any measure of success (Fig. 7).

4 Discussion: Relationships and Consequences

Hunicke et al. (2004) defined the relationship between the parts of MDA as follows: "Each component of the MDA framework can be thought of as a 'lens' or a 'view' of the game—separate, but causally linked… From the designer's perspective, the mechanics give rise to dynamic system behavior, which in turn leads to particular aesthetic experiences. From the player's perspective, aesthetics set the tone, which is born out in observable dynamics and eventually, operable mechanics." Apart from the terminology and some over-simplifications we already addressed, not only do we agree with this sequencing of both perspectives—the DDE framework also supports that explanation by simply replacing the old terms with new ones. However, underlying the entire premise there is still a potential for misconception because that sequencing does *not* describe a linear process. Instead, everything in the DDE design sequence happens in parallel.

From the designer's perspective, this nonlinear sequence is an iterative process, where each decision on every level can alter the whole game. From the player's perspective, we can further see this nonlinearity play out in the experience of a game. The player senses the game world *and* analyzes the underlying mechanics at the same time, while confronting and experiencing every aspect of the design. In practice, the designers are also a game's first players—long before the game is finalized.

When an implemented design is played, thus creating **Dynamics**, the design loses its static character and becomes at least partly driven by its mechanics: It becomes an agent of its own rules. If the designer thinks of this agent as a character, as the **Antagonist** of the **Player-Subject**, that perspective allows to comprehend the experience of playing a game as a single narrative (Walk 2016). As there are certain expectations players have about narratives, good designers will anticipate those expectations when creating a narrative experience. If the **Antagonist** of a narrative does not fulfill that role, the narrative will be less interesting than it might have been, and it may even fall apart.

In talking about narratives we do not only mean any embedded story that a narrative designer creates for a game. Of course, any embedded narrative should

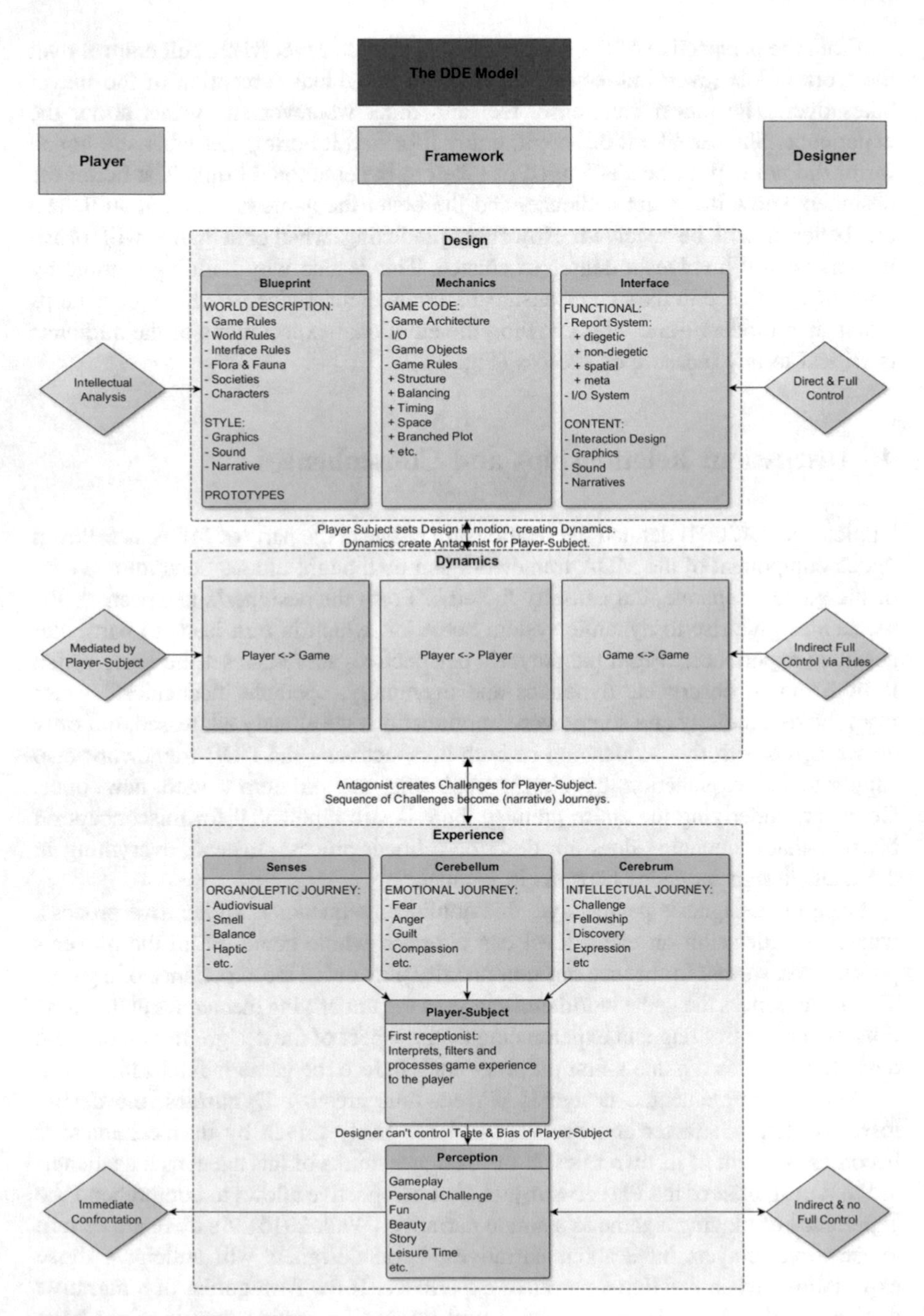

Fig. 7 The complete DDE framework

support and reinforce the complete experience—including the decision whether a narrative will be integrated at all. However, if the different journeys building the overall perception journey do not harmonize, the player will experience what Clint Hocking called "Ludonarrative dissonance" (Hocking 2007).

4.1 The Ludo-Narrative Disconnect and DDE

One story is the *embedded* story, created by the narrative designer; the other is the *emergent* story (or narrative journey) created by the sequence of challenges and other sensations emerging from the game dynamics (Grave 2015). Emergence in games occurs when the rules of the game system define both the challenges and the tools players can use to solve these challenges—without pre-defining solutions. Emergence is thus usually supported by procedural generation or re-combination of game content, rules and sometimes even whole levels. Help pages for emergent games usually have strategy guides as opposed to walkthroughs, rules of thumb and general tricks rather than step-by-step solutions (Juul 2002). The emergent, unanticipated gameplay of such games creates a unique journey for the **Player-Subject**; it is only repeatable if someone repeated the exact same sequence of orders to the game system at exactly the same time—and in case of randomly generated procedural contents it may not stay repeatable at all. As a result the **Player-Subject** perceives himself or herself as the hero of this emergent journey, which is inevitably understood *as* a narrative.

Both the embedded story and the player narrative have their dramatic arcs; both have their emotional content and sequence. However, if the two journeys fail to connect, the entire experience suffers from a weak, inconsistent **Antagonist** and the game will not deliver its full potential to the **Player-Subject**. Probably, both stories will even be perceived as weak if they might have been good stories when viewed separately. The authors think that it is here where most game stories fail. What narrative designers usually lack is not knowledge about storytelling or dramaturgy, but knowledge about the game's emergent **Antagonist**.

Just as the **Antagonist** in every narrative medium is responsible for the set of challenges the hero has to overcome, the **Antagonist** that grows out of the game's **Design** and **Dynamics** is—in confrontation with the **Player-Subject**—responsible for the obstacles the player will encounter. If the game is not a worthy **Antagonist**, the **Player-Subject** will stop playing or continue playing but be underwhelmed or even bored.

The DDE framework supports the notion of conflict from abstract to literal, whereas MDA does not mention it at all. It is here where we also believe that Winn's DPE model creates categories not supporting the design process. Winn wants to create a distinct framework for serious games by adding the "Learning" layer. Consequently, he has to make additional distinctions such as storytelling, gameplay and User Experience—as if learning was not the inevitable result of all those, or learning and experience are somehow separable entities.

The danger Winn evokes is that of a segmented design process—like the one between story and mechanics, only subdivided into four instead of two separate design instances: pedagogic content, narration, mechanics and user interface. Thus, the same inconsistent design principle currently leading to an endless stream of weak or failed narrative game experiences is multiplied to create pedagogic content on top. However, we do not claim that Winn is proposing this separation, but he also does not give an explicit warning. Even allowing for differences in terminology, a well-defined design framework should prevent such misconceptions, as they result in flawed designs.

The nonlinear relationship between the DDE framework's three categories means they are *not* to be interpreted as production stages. Rather, they are what must be understood in order to set up an effective iterative process for creating a game. As developers usually are the first players of their game, they will iterate the design until the dynamics produce the desired effects. They also are the first ones to perceive the game and to get an idea of the journey it will produce for the audience. With each iteration, every line of code and every asset can be adjusted to alter the effect the game has as an experience.

Iterative production cycles today should be a standard procedure. However, in our experience, they are the first thing to be abandoned when time or money is running short. Unfortunately, without iteration one *cannot* develop a good game. A thousand-piece jigsaw puzzle thrown in the air is never going to put itself together, and most games have considerably more than a thousand pieces.

5 Conclusion

The **Design, Dynamics, Experience (DDE) framework** is an attempt to overcome several weaknesses of the well-established MDA framework. The subcategories of the DDE framework, which in turn formed the three new main categories, are based on what actually needs to be produced during the design and development of a computer or video game as well as what the role of any produced asset will be during the production process or game experience.

We also took a closer look at the player's perspective, defining her experience as a journey. This shows the common denominator between the player's experience and a possible game narrative—which hopefully provides future game designers with a better understanding of the role of narratives in games. Consequently, DDE suggests that game designers employ an experience-oriented (as opposed to functionality-oriented) design process.

We also integrated Sicart's concept of the **Player-Subject** into the DDE framework to improve the understanding of the player's experience. Finally, we introduced the term **Antagonist** in which all challenges and narratives of a game come together as a single entity. Thus, DDE treats narrative design as an integral part of any design process for computer or video games, from the first day on.

DDE allows designers to assess and re-assess the value of the story within the overall framework of game development. Almost every substructure of the DDE framework has a narrative connotation. In terms of production, DDE endorses a highly iterative production process, in which everyone adopts the roles of designer *and* player in order to raise the bar of group awareness within the development team.

The DDE framework is a work in progress. So far, it serves the purpose of laying out a stable foundation for game design. Inevitably, the framework will undergo future changes. Some of the terminology may change, some definitions will receive additional attention—and there are still areas, which deserve being explored deeper.

References

Chou, Y. (March 1st, 2014). Octalysis: Complete Gamification Framework. Available from http://www.yukaichou.com/gamification-examples/octalysis-complete-gamification-framework

Crysler, C.G., Cairns, S. & Heynen, H. (2012). *The SAGE Handbook of Architectural Theory.*

Grave, G. (May 7th, 2015). Emergent narratives in games. Available from https://multiversenarratives.com/2015/05/07/emergent-narratives-in-games/

Herrmanny, K. & Schmidt, R., (2014). Ein Vorgehensmodell zur Entwicklung von Gameful Design für Unternehmen. In Butz, A., Koch, M. & Schlichter, J. (eds.), *Proc. of Mensch & Computer 2014* (pp. 369-377). Berlin: De Gruyter Oldenbourg.

Hocking, C. (October 7th, 2007). Ludonarrative Dissonance in Bioshock. The problem of what the game is about. Available from http://clicknothing.typepad.com/click_nothing/2007/10/ludonarrative-d.html

Hunicke, R., Leblanc, M. & Zubek, R. (2004). MDA: A Formal Approach to Game Design and Game Research. In *Proceedings of the AAAI Workshop on Challenges in Game AI.* Available from http://www.cs.northwestern.edu/~hunicke/MDA.pdf

Jilménez, S. (2014). Gamification Model Canvas. Available from http://www.gameonlab.com/canvas/

Juul, J. (2002): The Open and the Closed: Game of emergence and games of progression. In *Computer Games and Digital Cultures Conference Proceedings*, edited by Frans Mäyrä, 323–329. Tampere University Press. Available from http://www.jesperjuul.net/text/openandtheclosed.html

Kumar, J. & Herger, M. (2013). *Gamification at Work. Designing Engaging Business Software* (1st edition). Available from https://www.interaction-design.org/literature/book/gamification-at-work-designing-engaging-business-software

Lantz, F. (September 14th, 2015). MDA. Available from http://gamedesignadvance.com/?p=2995

Polansky, L. (2015). On Genre and the Ludic Device. Available from http://sufficientlyhuman.com/archives/1008

Osterwalder, A. & Pigneur, Y. (2011). Business model generation. Ein Handbuch für Visionäre, Spielveränderer und Herausforderer. Frankfurt: Campus-Verlag.

Ralph, P. & Monu, K. (2014). A Working Theory of Game Design. Mechanics, Technology, Dynamics, Aesthetics & Narratives. Available from http://www.firstpersonscholar.com/a-working-theory-of-game-design/

Schell, J. (2008). *The Art of Game Design*. Morgan Kaufmann.

Sicart, M. (2009). The Ethics of Computer Games. Cambridge/London: MIT Press.

Silveira Duarte, L.C. (February 3rd, 2015). Revisiting the MDA framework. Available from http://www.gamasutra.com/blogs/LuizClaudioSilveiraDuarte/20150203/233487/Revisiting_the_MDA_framework.php

Stonehouse, A. (February 27th, 2014). User interface design in video games. Available from http://www.gamasutra.com/blogs/AnthonyStonehouse/20140227/211823/User_interface_design_in_video_games.php

Walk, W. (August 8th, 2016). The Ethical Avatar. Available from http://www.gamasutra.com/blogs/WolfgangWalk/20160808/278701/The_Ethical_Avatar.php

Werbach, K. & Hunter, D. (2012). *For the win. How game thinking can revolutionize your business*. Philadelphia: Wharton Digital Press.

Winn, B. M. (2008). The Design, Play, and Experience Framework. In Ferdig, R. E. (ed), *Handbook of Research on Effective Electronic Gaming in Education*, Volume III, Chapter LVIII. Kent: Kent State University.

Author Biographies

Wolfgang Walk studied German literature and philosophy while working on his first computer games in the early 1990s. In 1995, he joined Blue Byte as a project lead and writer. Collaborating on a Blue Byte game in 1996 with freelance writer, Mark Barrett led to a long-lasting collaboration on many projects in business and private.

Working as a freelance producer and designer since 2005, Walk teaches in the field of interactive storytelling at numerous universities in Germany, and speaks frequently at conferences. He has also consulted for Fraunhofer Institute, SWR, Bigpoint, and many others, and worked on games for companies from the Netherlands to Italy to Russia.

Increasing dissatisfaction with the developmental state and art of interactive storytelling—especially the industry's tendency to believe that mimicking film techniques is sufficient—has led to years of private research and discussions with other game writers and designers.

Walk is working with Mark Barrett on a book about the narratization of games and regularly shares his design insights on his blog *Blindband*.

Daniel Görlich is a professor for virtual reality, game development, and film informatics at SRH University Heidelberg, Germany's biggest and oldest private campus university.

After completing his diploma in computer science and psychology, he worked as a usability consultant, senior consultant, and team leader of HMI at the Center for Human–Machine Interaction (ZMMI) at the Technical University of Kaiserslautern and the German Research Center for Artificial Intelligence (DFKI). His research focused on future human–machine interaction technologies. For his work about automatic software generation for ambient intelligence environments, Görlich received his Ph.D. in mechanical engineering in 2009. He was then offered a management position at the Fraunhofer Institute for Experimental Software Engineering (IESE) in 2010, and a professorship at SRH University Heidelberg in 2012, where he currently heads the B.Sc. study course on Virtual Realities.

Originally, Görlich had learned programming by re-implementing famous video game classics back in the late 1980s. He prefers strategical games, construction and management simulations, and SciFi action role-playing games. His primary research interest concerns the convergence of modern media, especially interactive media such as games, with future interaction technologies and techniques.

Mark Barrett studied fiction writing at the University of Iowa and then worked as a screenwriter in Los Angeles before turning to games as a narrative designer in the early 1990s. In 1995, he joined Chris Crawford and others on the CompuServe GameDev forum and began decades of wrestling with interactive storytelling in theory and practice, including contributing to the conversation with GDC roundtables on Emotional Involvement from 2000 to 2002.

Since 1996, Mark has collaborated with Wolfgang Walk on more than 15 games, on the script of a rock opera set in a game development studio, and on the development of a design framework which includes narrative design as an integral part of the design process. This framework is intended not only as a comprehensive theoretical approach, but also as a practical guide to understanding and implementing interactive storytelling as an art form.

DDE is part of the result of that research, and a cornerstone of a planned book on the subject. Articles on interactive storytelling and moderator's reports from Mark's GDC roundtables can be found on his eclectic *Ditchwalk* blog.

Procedural Synthesis of Gunshot Sounds Based on Physically Motivated Models

Hüseyin Hacıhabiboğlu

Abstract Generation of content for games is one of the major bottlenecks in terms of the effort required and the resources to be committed. A typical AAA game contains tens of thousands of sound files as audio assets, which include spoken dialogue as well as sound effects. Procedural content generation (PCG) provides a cost-effective alternative to recording these sounds in the studio or in the field. While some sound effects can be recorded with comparatively easy, given the necessary time, effort, and resources, some others such as gunshot sounds are not easy to record. Since many games and simulations incorporate firearms, parametric sound synthesis, which is essentially a PCG technique can be used to alleviate the need to record gunshot sounds. This chapter describes a physically motivated parametric gunshot sound synthesis model. The model is based on a deconstruction of the gunshot sound event into its constituent parts and uses parameters such as the barrel length, bullet type, and muzzle velocity to synthesise the sounds of different firearms. A subjective evaluation, which investigates the perceptual relevance of the proposed model, is also presented.

1 Introduction

Sound effects are essential components of computer games. While it is difficult to estimate the average number of audio assets included in a game, it is reasonable to expect that potentially tens of thousands of sounds could be needed in an AAA title. For example, Battlefield 4 (Electronic Arts 2016) with its expansion packs has in excess of 78,000 sound files (BF4 audio files 2014). Considering how hard it is to record so many sounds, procedural audio generation emerges as a possible solution. Many different sounds can be synthesised using procedural means instead of being recorded, and this would reduce both the size of the game and the substantial effort

H. Hacıhabiboğlu (✉)
Graduate School of Informatics, Middle East Technical University (METU), Ankara, Turkey
e-mail: hhuseyin@metu.edu.tr

© Springer International Publishing AG 2017

O. Korn and N. Lee (eds.), *Game Dynamics*,
DOI 10.1007/978-3-319-53088-8_4

required for the task. Besides, some sounds such as wind or rain are very hard if not impossible to record, due to physical constraints. Procedural sound generation provides a viable solution for such cases too.

Shooters remain one of the most popular video game genres constituting 24.5% of the video games sold in the USA in 2015 (Statista Inc. 2016). Shooters such as the *Battlefield* series or the serious game *America's Army* (US Department of Defense 2002) require a variety of different weapon sounds. Game mechanics that involves shooting requires the playback of the sounds of a variety of firearms. For this reason, recording or generation of high quality weapon sounds is essential. As an example, Battlefield 4, praised for its sound design, has over 3900 recorded sound effects for weapons.

The problem with weapon sounds is even more profound with games which incorporate procedural generation. The games *Borderlands* (Gearbox Software 2009) and *Stack Gun Heroes* (Team Stack Studios 2015) are good examples, which allow the parametric generation of a huge number of weapons, all having different properties. In such games, sounds of weapons, which typically will not exist in real life, also have to be generated.

The aim of this chapter is to present a physically motivated, generalised procedural gunshot sound synthesis model, which can be used to generate the sounds of a wide variety of classes and types of firearms with subsonic and supersonic projectiles. The presented synthesis model is simple, flexible, and modular. Thus, it can be applied as part of a larger chain of procedural sound generation algorithms. The perception of the synthetic sounds obtained from the model are compared with real recordings in a subjective experiment and the results indicate that procedurally generated gunshot sounds can possibly replace real gunshot recordings without substantially reducing the perceived realism.

2 Recording Gunshot Sounds

Generating a large and generic database of recordings of gunshot sounds for use in games, simulations, and other virtual reality applications may seem as a reasonable action to take. However, there are several drawbacks. First, legal difficulties may prohibit anyone interested in developing such a database from possessing an arsenal of different firearms especially in countries where strict gun control laws are in place.

Leaving this difficulty aside, several other technical issues make it difficult to record firearm sounds. The first difficulty is the sound pressure levels (SPL) observed with gunshot sounds. Depending on the gun and ammunition properties, peak signal level can easily exceed 150 dB SPL. Such a level is usually not handled well by general-purpose recording microphones used by sound effects artists and is likely to result in saturation and clipping rendering the original signal unrecoverable. The diaphragm size of the employed microphone, the maximum sound pressure level that it can handle, and its sensitivity are thus very important in obtaining a recording that is free of artefacts (Rasmussen et al. 2009).

Another important difficulty is the inclusion of the acoustics of the environment in the recording. If the sound of the weapon is recorded anywhere other than an anechoic chamber, the reflections and the reverberation pertaining to the environment cannot be excluded from the recorded signal. If the same gun can be used in different environments in the game world, such as in a warehouse, a forest, a city canyon, or an open field, the mismatch between the acoustics of the recording venue and that of the virtual environment in the game would negatively affect its realism. Consider, for example, the case where the recordings are made in an outdoor shooting range and the game world is primarily designed as indoors.

Care has to be taken at every level of the audio recording chain while recording gunshots. An important difficulty, which is often overlooked, is the sampling rate. As will be discussed below, components of gunshot sounds have very short durations and thus gunshots will in general have very wide frequency ranges. A sufficiently high sampling rate should be used to capture a good representation of the gunshot.

Another aspect of gunshots is that they are highly directional (Beck et al. 2011). Largest sound pressure levels are observed at the boreline direction and the sound level decreases progressively as the recording direction moves away from the boreline direction. Behind the firing position, the body of the person firing the gun shadows the sound wave. For supersonic projectiles, the composition of the gunshot sound also changes with direction. The shock wave may or may not be observed based on the recording position. This makes it necessary to obtain multiple simultaneous recordings around the shooting position, which is impractical most of the time. For the reasons stated above, gunshot recordings are in general not suitable for use in serious games where good levels of authenticity and realism are required.

3 Synthesis of Gunshot Sounds

Gunshot sounds experienced in real life and as used in games and movies are profoundly different. The sound of a gunshot in a game acts as an expressive device to elevate the excitement, and possesses diegetic and expectational relevance but not necessarily realism. Simulations, on the other hand, require authentic or near-authentic gunshot sounds due to the training requirements, and the need to satisfy the expectations of the users who are already accustomed to how a specific firearm sounds like.

3.1 Properties of Gunshots

Sound resulting from a gunshot is a transient, high-energy signal, which has well-defined properties (Carlucci and Jacobson 2013). These properties depend not only on the physical features of the firearm and the ammunition which generates the sound, but also on the environment in which the gun is fired and the direction of the microphone with respect to the firearm (Maher 2009). More specifically, a gunshot sound consists of the muzzle blast and the shock wave. Muzzle blast is localised at the muzzle and occurs due to the explosion of the propellant in the bullet. Shock waves (i.e. also known as sonic boom or N-wave) occur only for bullets or projectiles, which have supersonic speeds. These waves will have distinct propagation patterns following the projectile and spectra with a wide bandwidth due to the discontinuity they possess.

The sound generated by a firearm is also affected by (i) the observation distance, (ii) the direction of observation, and (iii) the environment. The primary effect of distance is by attenuating the level, and absorbing the high-frequency content due to thermal relaxation process (Kuttruff 2009). The direction of observation is an important factor in whether or not the shock wave will be audible and in the relative level of the muzzle blast sound with respect to its front direction (Maher and Shaw 2010). If the gunshot occurs in open air with no nearby reflectors apart from the ground, the effect of the environment will be limited by a single ground reflection. Any other environmental setting such as a room or a forest will also add reverberation. A realistic presentation of a gunshot should thus incorporate the sonic properties of the gun as well as the properties of the environment.

3.2 *Parametric Synthesis of Gunshots*

Parametric sound synthesis is a very active field of research and many different models have previously been proposed for different purposes (Farnell 2010). For example, there are models for footsteps (Nordahl et al. 2011), birdcalls (Kahrs and Avanzini 2001), and vehicle sounds (Cascone et al. 2005). Research into affective synthesis of speech is also an active field (Schröder 2009) that will soon make it possible to tie conversational agents to NPCs and to create games with dynamically evolving scripts.

A crucial component in the procedural generation of a sound is the availability of computationally tractable models of the processes, which are involved in the generation of that sound in real life. Such models can be physical, perceptual, or statistical.

The sound generated by a gunshot is above all a physical phenomenon. There exist well established but approximate theoretical models for different portions of a gunshot sound and the properties of the acoustical environment discussed above. Regardless of the properties of the firearm used, muzzle blast has a distinctive shape with a sharp onset followed by positive and negative phase portions, respectively (Beck et al. 2011). While this waveform is common across many different types of firearms, the durations of the positive and the negative phases as well as the peak pressure at the onset will depend on the calibre, the form factor, and the length of the projectile, as well as the length of the barrel. The peak pressure also depends on the direction of observation around the firearm.

The pressure signature of the shock wave due to a supersonic bullet depends mainly on the dimensions and the speed of the bullet. While an exact analytical solution of the pressure signature is not feasible, simplified models can be used to predict the shape of the N-wave (Beck et al. 2011).

The effects due to distance can be simulated by using two distinct attenuation models for the muzzle blast and the shock wave. The high-frequency attenuation can be simulated using specially designed air absorption filters (Huopaniemi et al. 1997). In addition, the relative level of the shock wave will depend on the observation angle and should be simulated accordingly. There are several different artificial reverberation algorithms, which can be used to simulate different environments, e.g. an enclosure (De Sena et al. 2015) or a forest (Spratt and Abel 2008). In open field, the only environmental effect is the ground reflection, which can be simulated by adding to the original signal a delayed and attenuated version of itself.

Physical models of firearm acoustics as well as a simple parametric gunshot sound generator are described in the following section.

4 Physically Motivated Synthesis of Gunshot Sounds

A gun is a single-stroke steam engine designed to throw a bullet in a desired direction with a very high speed. The bullet forms the tip of the cartridge, which also consists of the casing, the propellant, and the primer. The propellant is typically a combination of some oxygen and chemicals which when ignited will produce rapidly expanding gas to propel the bullet through the barrel and out from the muzzle. Figure 1 shows an ultra-high-speed photograph of a bullet exiting the muzzle of Smith and Wesson 686.38 Special revolver pistol.

Muzzle blast is clearly visible in the photograph. Parts of an automatic handgun and ammunition are shown together with the parameters that affect the pressure

Fig. 1 Ultra-high speed photograph of bullet fired from a SW revolver photographed with an air-gap flash by Niels Noordhoek, https://www.scopus.com/authid/detail.uri?authorId=6603387174 (CC BY-SA 3.0)

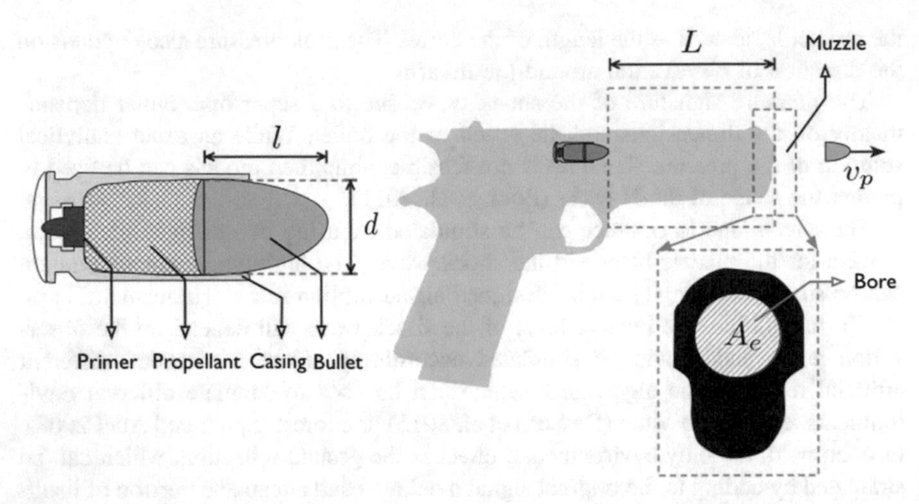

Fig. 2 Diagrams of a projectile and a M1911 handgun. Several different parameters used in the synthesis model are also shown

signal generated due to the gunshot in Fig. 2. Sounds generated during a gunshot can be separated into three elements: (i) ballistic sounds such as muzzle blast and shock wave, (ii) mechanical action sounds, and (iii) environment effects such as reflections and reverberation. The first group is what is known as the gunshot and consists primarily of the muzzle blast and shock wave. The second group consists of sounds generated by the internal mechanics of the firearm before, during, and after the gun is fired. The third group includes the effects of the environment such as reverberation and attenuation of high frequencies.

This chapter is concerned mainly with the first and third group of sounds, which will be discussed below. The second group consists of sounds such as the trigger action and bullet loading and in general has a much lower signal level rendering them practically inaudible sufficiently far away from the firearm location.

4.1 *Muzzle Blast*

Muzzle blast is the explosive sound which occurs as the expanding gases rapidly escape from the muzzle. The associated pressure signal has a distinct shape which depends on many factors. A generic muzzle blast waveform is given in Fig. 3. It may be observed that the waveform has a very sharp onset immediately reaching a peak pressure level, $\widehat{P}_m$, followed by an exponentially decaying positive pressure phase and a negative pressure phase at which the pressure level falls below the ambient pressure, eventually reaching the ambient pressure level, $P_0 \approx 101\,\text{kPa}$.

The muzzle blast waveform is a typical example of a Friedlander wave (Friedlander 1946) and can be represented using the following parametric form:

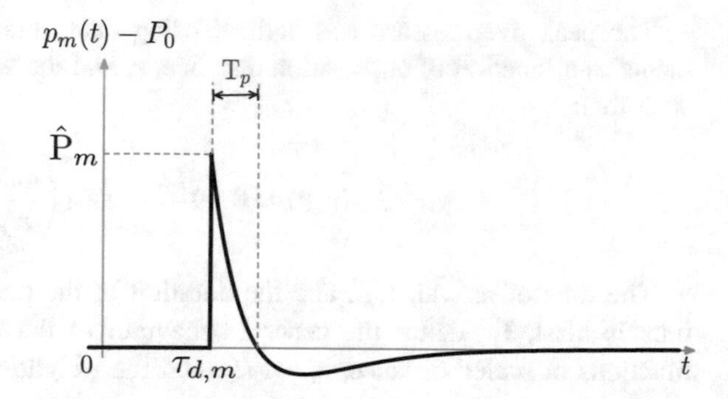

Fig. 3 An ideal muzzle blast waveform showing the pressure due to the gunshot as a function of time. $\widehat{P}$ is the peak overpressure, $\tau_{d,m}$ is the delay, and T_p is the positive phase duration

$$p_{s(t)} - P_0 = \widehat{P}_m(r, \theta)\left(1 - \frac{t - \tau_{d,m}}{T_p}\right)e^{-\frac{t-\tau_{d,m}}{T_p}} \tag{1}$$

where P_0 is the ambient pressure, $\widehat{P}_m(r, \theta)$ is the peak overpressure at a given distance r and a direction θ with respect to the boreline, $\tau_{d,m}$ is the delay of arrival of the blast wave at the observation position, and T_p is the duration of the positive phase. There are several different studies investigating the properties of gunshot sounds. An adaptation of the model described in Fansler et al. (1993) is used in this chapter for calculating T_p and $\tau_{d,m}$.

The employed model is based on the concept of scaling length defined for the boreline direction of the gun as:

$$l_s \propto \sqrt{(\mathrm{d}E/\mathrm{d}t)/(cP_0)} \tag{2}$$

where $\mathrm{d}E/\mathrm{d}t$ is the *energy deposition rate* due to the blast and is a function of the propellant properties, peak muzzle pressure, speed of the propellant at exit, the Mach number at projectile ejection, *Me*[1], and the bore area of the barrel. These parameters are typically provided by weapon and ammunition manufacturers.

The directional properties of the muzzle blast are modelled by weighting the scaling length with respect to the angle of the observation point from the boreline direction, θ:

$$l_{sn}(\theta) = \left(\mu\cos\theta + \sqrt{1 - \mu^2\sin^2\theta}\right)l_s \tag{3}$$

where μ is the *momentum index* defined as the ratio of the sound pressure at the front and at the rear of the firing position.

---

[1]Mach number is the ratio of the projectile speed to the speed of sound in air under the same atmospheric conditions, i.e. $M = v_p/c$

The peak overpressure was defined using data obtained from actual measurements as a function of observation distance, r, and the weighted scaling length, l_{sn}, such that:

$$\widehat{P}_m(r,\theta) = 0.89\frac{l_{sn}}{r} + 1.61\left(\frac{l_{sn}}{r}\right)^2 \quad (4)$$

The time of arrival, $\tau_{s,m}$, and the duration of the positive phase portion of the muzzle blast, T_p, define the general behaviour of the wave and are modelled as functions of scaled distance, $r_s = r/l_s$, and the weighted scaling length, l_{sn}:

$$\tau_{d,m} = \frac{l_{sn}}{r}[X(r_s) - 0.52\ln(2X(r_s) + 2r_s) - 0.56] \quad (5)$$

where

$$X(r_s) = \sqrt{r_s^2 + 1.04r_s + 1.88}. \quad (6)$$

The model in Fansler et al. (1993) defines the positive phase duration differently for $r_s < 50$ and $r_s \geq 50$ such that:

$$T_p = \begin{cases} \frac{r_s}{c} - \tau_{d,m} + G, & r < 50 \\ \frac{l_{sn}}{c}[2.99\sqrt{\ln(33199r_s)} - 8.534 + G], & r \geq 50 \end{cases} \quad (7)$$

where G is a correction parameter obtained from measurements.

Figure 4 shows the muzzle blast signals calculated for different pistols used in this chapter. It may be observed that while the shapes of the signals are similar, the peak overpressure and positive phase durations are different.

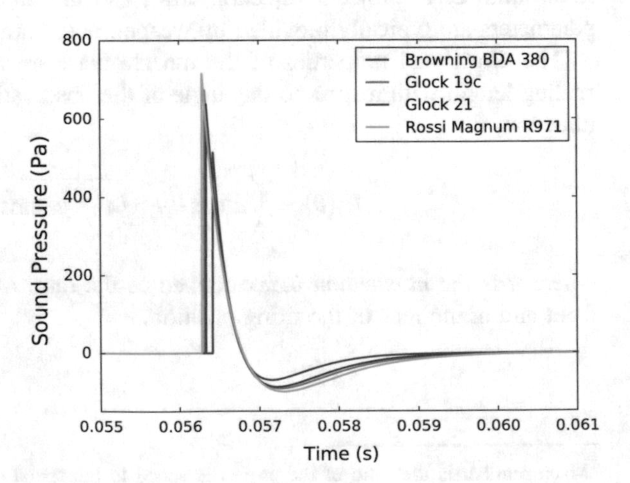

Fig. 4 Muzzle blast signal for different pistols and ammunition obtained using the described model for a distance of 20 m away from the firing position

4.2 Shock Wave

Projectiles that travel faster than sound generate pressure waves that have very sharp onsets and offsets. These waveforms also have strong positive and negative phase parts and a distinctive shape. They are called an N-waves and the accompanying sound is called a sonic boom.

The production of the shock wave and its propagation characteristics is above all a function of the projectile speed. For projectiles travelling at speeds higher than the speed of sound (i.e. supersonic speeds), the shock wave will travel outwards on the conical surface trailing the projectile. This is known as the Mach cone. Half of the opening angle of the cone is called the Mach angle, θ_M and depends on the Mach number M_e such that:

$$\theta_M = \sin^{-1} M_e^{-1} \tag{8}$$

Mach cone and Mach angle are shown in Fig. 5. It should be noted in order for the shock wave to be observed, the observation point needs to be outside the Mach cone. For example, the person firing the gun will not hear it.

Figure 6 shows the ideal waveform due to an N-wave. The wave is symmetric about $\tau_{d,s} + T_d/2$ so it can be defined by four distinct parameters: (i) rise time, t_r, (ii) total duration of the wave, T_d, (iii) peak overpressure, $\widehat{P}_s$, and (iv) the time delay of arrival, $\tau_{d,s}$. The time delay of arrival will be calculated in the next section. The other parameters based on earlier work (Whitham 1952; Stoughton 1997) are described below.

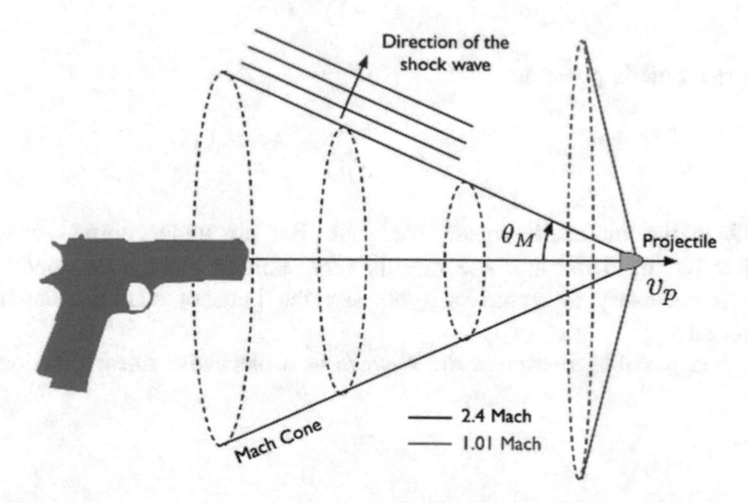

Fig. 5 Mach cone due to a projectile travelling at supersonic speeds. Two Mach cones for projectile speeds of 2.4 and 1.01 Mach are shown (adapted from Maher and Shaw 2008)

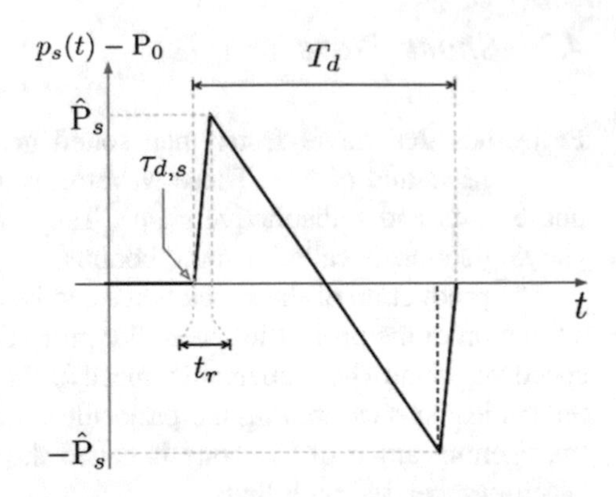

Fig. 6 Waveform of an ideal sonic boom showing the pressure due to a supersonic gunshot as a function of time. $\widehat{P}_s$ is the peak overpressure, $\tau_{d,s}$ is the time delay, t_r is the rise time, and T_d is the total duration of the N-wave

The peak overpressure is the maximum pressure level that the shock wave will achieve:

$$\widehat{P}_s = 0.53 \frac{\left(M_e^2 - 1\right)^{1/8} d}{L_m^{3/8} l^{1/4}} \tag{9}$$

where d is the diameter of the projectile (in m), l is the length of the projectile (in m), and L_m is the miss distance (i.e. the geometric distance of the observation point to the trajectory of the projectile).

The total duration of the N-wave is given as:

$$T_d = \frac{1.82 M_e L_m^{1/4} d}{c\left(M_e^2 - 1\right)^{3/8} l^{1/4}} \tag{10}$$

and the rise time is given as:

$$t_r = \frac{\lambda P_0}{c \widehat{P}_s} \tag{11}$$

where λ is the molecular mean free path. For air under normal conditions $\lambda \approx 6.8 \times 10^{-8}$m. Typically, rise time is very short in the microseconds range, making it necessary to record or synthesise the gunshot sounds using a high sampling rate.

It is then possible to express the N-wave as a piecewise linear function, such that:

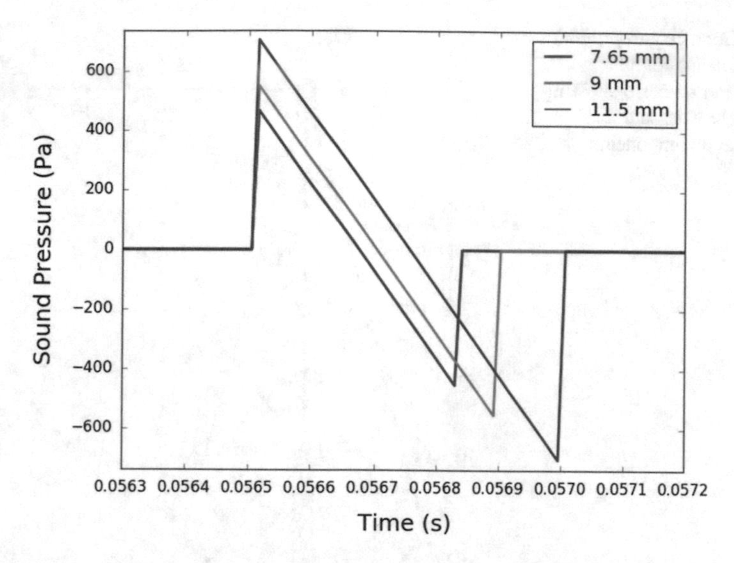

Fig. 7 Shock wave signals calculated using the described model for different projectile diameters calculated at a distance of 20 m for a theoretical gun with a barrel length of 10 cm, projectile length of 1 cm, exit pressure of 2000 kg/cm^2, and an exit velocity of 380 m/s for different projectile diameters

$$p_s(t) = \begin{cases} 0, \quad t \le \tau_{d,s} \text{ and } t \ge \tau_{d,s} + T_d \\ \widehat{P}_s \frac{t-\tau_{d,s}}{t_r}, \quad \tau_{d,s} < t \le \tau_{d,s} + t_r \\ \widehat{P}_s\left(1 - 2\frac{t-\tau_{d,s}-t_r}{T_d-2t_r}\right), \quad \tau_{d,s} + t_r < t \le \tau_{d,s} + T_d - t_r \\ \widehat{P}_s\left(\frac{t-\tau_{d,s}-T_d+t_r}{t_r} - 1\right), \quad \tau_{d,s} + T_d - t_r < t \le \tau_{d,s} + T_d \end{cases} \tag{12}$$

This parametric form also allows a straightforward way to obtain the pressure value at any point in time, simplifying the sound synthesis process. Figure 7 shows N-waves calculated using the described parametric model for different projectile diameters.

4.3 Geometry of Propagation

The gunshot sound signal under anechoic conditions can be expressed as a combination of the sound signals due to the muzzle blast and the shock wave, if the latter is present. The geometry of propagation for these two components is different.

The geometry of the gunshot sound propagation scenario is shown in Fig. 8. The two components will travel for different durations and arrive at the microphone at different incidence angles. The time delay of arrival of the muzzle blast was discussed previously. The time delay of arrival for the shock wave can be calculated as:

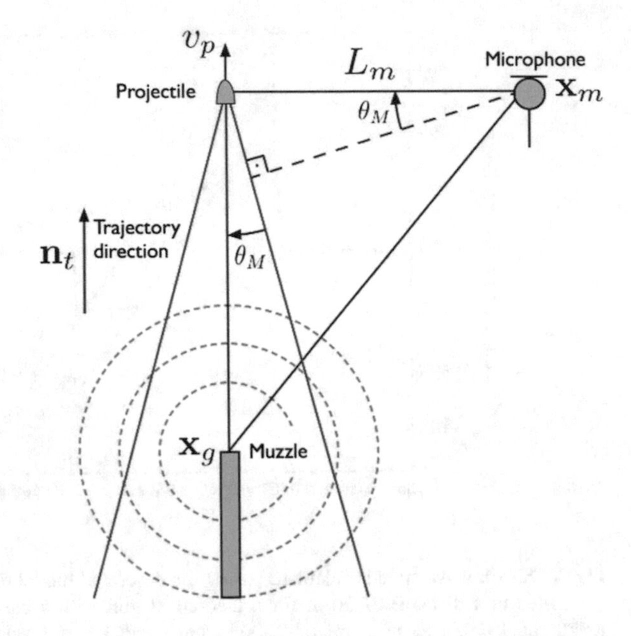

Fig. 8 Geometry (*top view*) of the gunshot sound propagation scenario showing the muzzle blast and the shock wave components

$$\tau_{d,s} = \left(\langle \mathbf{x}_m - \mathbf{x}_g, \mathbf{n}_t \rangle + L_m \cos\theta_m\right)/c \tag{13}$$

where $\langle\cdot,\cdot\rangle$ denotes the inner product of two vectors, $\mathbf{x}_m$ is the microphone position, $\mathbf{x}_g$ is the muzzle position, $\mathbf{n}_t$ is the unit vector in the direction of the trajectory, θ_m is the Mach angle, and L_m is the miss distance.

4.4 Effects of the Environment

In the absence of atmospheric effects such as wind or temperature gradients, three main environmental effects on gunshot sounds can be considered. The components of the gunshot sound will be reflected, their high-frequency components will be attenuated at greater distances, and in enclosures such as rooms, gunshot sound will be reverberated.

In an open field, the sound recorded at the observer position will include a reflection incident from the ground in addition to the original gunshot sound. It should be noted that as with the direct path from the gun to the observation point, the ground reflection paths will also be different for the muzzle blast and the shock wave. The delays of the two reflections are given as:

$$\tau_{mb,g} = \left\|\mathbf{x}_g - 2h_g\mathbf{u}_z - \mathbf{x}_m\right\|/c \tag{14}$$

$$\tau_{sb,g} = \|\mathbf{x}_{sb} - 2h_{sb}\mathbf{u}_z - \mathbf{x}_m\|/c \tag{15}$$

where $\mathbf{x}_{sb} = \mathbf{x}_g + L_m(\cot\theta_{m,g} - \tan\theta_M)\mathbf{n}_t$ is the shock wave source position, $\mathbf{u}_z$= [0 0 1] and h_g and h_{sb} are the height of the muzzle and the height of the shock wave source position from the ground, respectively.

The effect imposed by the ground on the sound wave depends on the surface properties as different ground surfaces (such as grass, concrete, or gravel) will have different absorption characteristics. Most surfaces, however, will absorb more of the high-frequency content, smoothing the original gunshot sound. Absorption characteristics tabulated elsewhere (Vorländer 2007) can be used to design digital filters (Huopaniemi et al. 1997) to simulate acoustical absorption.

Two other important environmental effects are attenuation and air absorption which cause the overall sound level to decrease with distance and the higher frequency components to be attenuated, respectively.

Pressure due to a spherical wave decreases with the radial distance. This is known as the *spherical spreading law* or the *1/r-law*. The employed firearm acoustics model already takes this attenuation into account. It is thus unnecessary to model these losses separately.

Other than attenuation, air absorbs a significant amount of high-frequency energy from a sound wave. The main reason for this is the physical process called *thermal relaxation* (Kuttruff 2009). A closed form expression for frequency-dependent air absorption as a function of humidity is given in International Organization for Standardization (1993), and it was shown (Huopaniemi et al. 1997) that this functional form can be used for designing simple filters for simulating air absorption. Figure 9 shows the air absorption and attenuation for different source distances.

The magnitude responses of fourth-order infinite impulse response (IIR) filters are also shown with broken lines. It may be observed that air absorption is mostly effective at higher frequencies and at larger source distances. In the context of gunshot sounds, this corresponds to a significant reduction in higher frequencies at large distances.

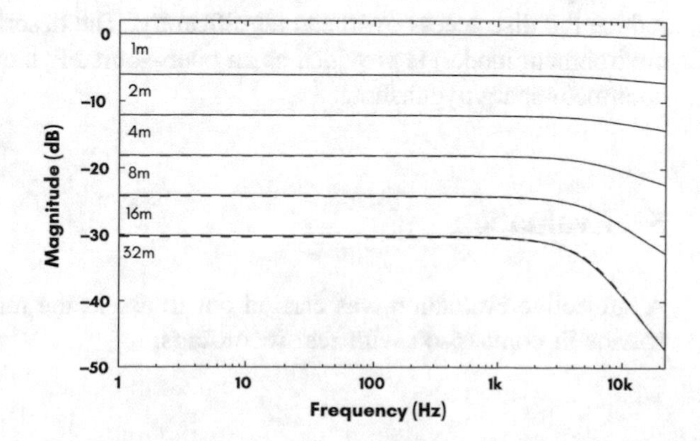

Fig. 9 Air attenuation and frequency-dependent air absorption. The *solid curves* show the physical model and the *broken curves* show the filters designed to emulate the given distances. It may be observed that larger distances correspond to higher absorption and lower levels for high frequencies

When a weapon is fired indoors, the acoustical effect of the enclosure will also have a profound effect on the generated sound. Environments with many reflecting and scattering objects such as rooms, streets, or forests are multipath environments. In such environments, first-order reflections of sound sources are observed together with higher-order reflections and reverberation. In order to simulate the effects of different environments, existing artificial reverberators can be used. While one of the most commonly used reverberation algorithms is the Jot's reverberator (Jot and Chaigne 1991), more recent algorithms such as scattering delay networks (SDN) (De Sena et al. 2015) and *treeverb* (Spratt and Abel 2008) are more suitable as they provide means to simulate position-related aspects of the gunshots more easily and provide environment-specific reverberation. Readers interested in artificial reverberator algorithms are referred to two recent review articles (Välimäki et al. 2012, 2016).

4.5 Synthesis Model

The synthesis model used in this chapter uses two sound synthesis modules: one for the muzzle blast and one for the shock wave. These modules use parameters from ballistic databases that can be bundled with a game, simulation, or a VR application. The positions of the listener (player) and the firearm can be fed to the model from the game at runtime for the muzzle blast and shock wave calculations. The simulation of the acoustics of the environment is handled in a second stage. Scene description including the listener and firearm positions can also be obtained at runtime from the game. Figure 10 shows the synthesis model.

It should be noted that just a small number of parameters is sufficient to synthesise the sounds of a wide variety of firearms. Since in an interactive environment positions of the shooter and the observation position are always changing, some parameters such as the delays of each component of the gunshot sound need to be refreshed at runtime. However, since gunshot events are likely to be sparse and also the calculation of the mentioned parameters is trivial, the computational cost would be negligible. The described algorithm, as with most other PCG methods, will also reduce the disc access overhead significantly. The described algorithm (except the environment model) is provided as an open-source Python module at https://github.com/metu-sparg/pygunshot.

5 Evaluation

A subjective evaluation was carried out to assess the realism of synthetic gunshot sounds in comparison with real recordings.

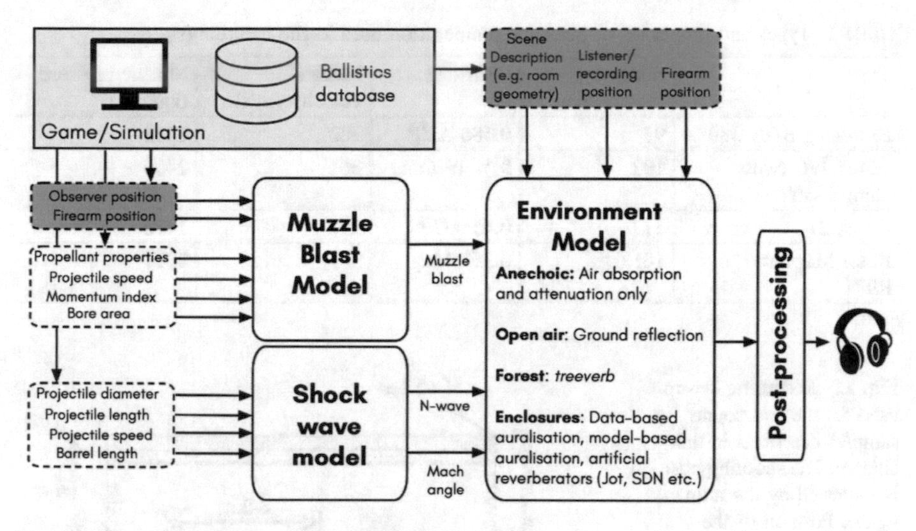

Fig. 10 Parametric gunshot sound synthesis model described in this chapter

5.1 Measurements

A set of measurements were carried out in the indoor pistol shooting range located at the head offices of Mechanical and Chemical Industry Corporation (MKEK) which is a major firearm and ammunition producer based in Ankara, Turkey. Three repeated recordings from four different pistols were made.[2] Table 1 gives details of the pistols and the ammunition used. It may be observed that two of the recorded pistols (Glock 19c and Rossi Magnum R971) nominally generate shock waves along with the muzzle blast. The shooting range has a shoebox geometry with the dimensions $W = 10.3$ m, $L = 26$ m and $H = 3.1$ m. The room is acoustically treated with porous rubber acoustic tiles on the side walls, and with tilted steel backstops covered by a thick rubber curtain at the end of the range. The reverberation time averaged across frequencies is $T_{30} \approx 0.68$ s. Figure 11 shows the recording geometry.

An omnidirectional microphone (Alctron M6) with a nominal maximum SPL value of 149 dB SPL connected to the microphone preamps of a MOTU 896 Mk3 sound card was used the record the gunshots. In order to increase its dynamic range to prevent clipping, the microphone was muffled with a tight latex cover. This resulted in a 15 dB improvement in the dynamic range allowing higher levels to be

[2]The recorded and synthetic gunshot sounds are made available at https://figshare.com/s/89d90977887b0bb8f54c.

Table 1 Types and properties of guns and ammunition used in the recordings

Type	Barrel length (mm)	Cartridge	Muzzle velocity (m/s)	Muzzle pressure (kg/cm²)
Browning BDA 380	97	0.380 ACP	300	1511
Glock 19C (with compensator)	102	9 × 19 mm	361	2460
Glock 21	117	0.45 ACP	291	1476
Rossi Magnum R971	101.6	0.357 Mag	388	2817

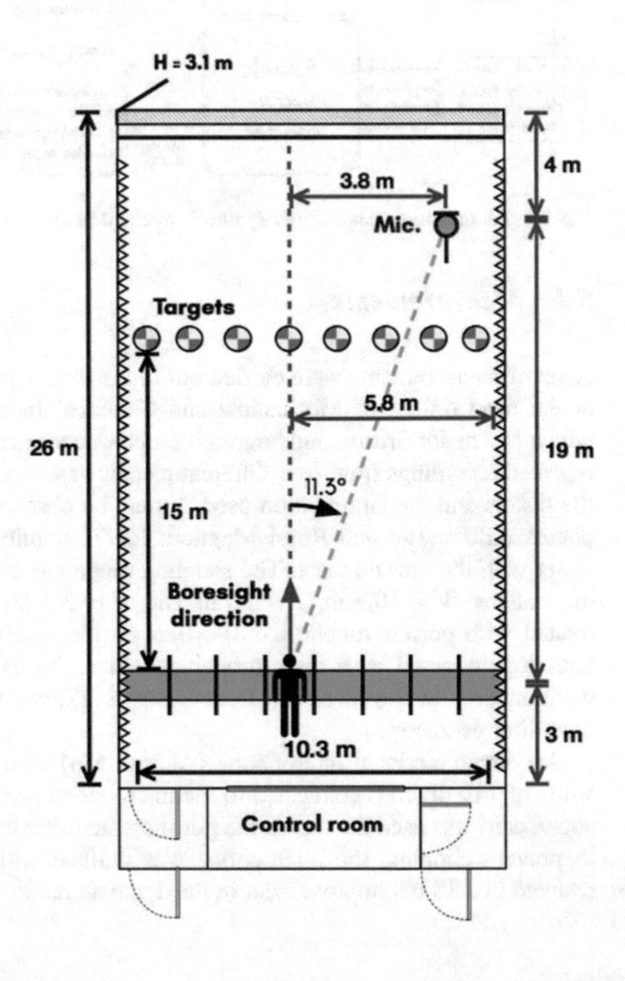

Fig. 11 Recording set-up used for the real sound samples described in this chapter. The shooting position is indicated by the human figure. Position of the microphone is also indicated

recorded without clipping. The microphone was positioned 19 m away from the firing line and 3.8 m away from the target at a height of 1.56 m. The sampling rate used in the recordings was 192 kHz.

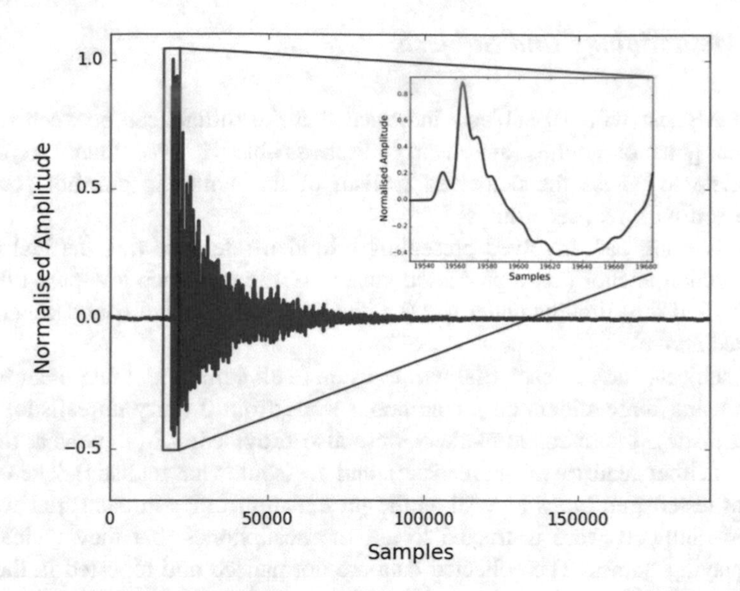

Fig. 12 Pressure signature of the Browning BDA 380 recorded in the indoor firing range. While the larger figure mainly shows the reverberation response of the room, the inset clearly shows that the direct path component has the Friedlander wave shape

An example sound signal from the recording of the Browning BDA 380 pistol is shown in Fig. 12. The Friedlander wave is clearly visible in the inset figure.

5.2 *Stimuli*

The sounds used in the subjective experiments were the recordings made in the indoor firing range as described above and samples synthesised using the algorithm described in this chapter.

The synthesis algorithm by itself does not incorporate the reverberation due to the recording environment. In order to add reverberation, an SDN-type reverberator was used. SDN allows the selection of the room size and absorption characteristics as well as the source and microphone positions. It was made sure that the simulated reverberation corresponded well to the acoustics of the firing range.

Three additional but trivial steps of processing were also applied on the synthetically generated sounds: Despite precautions the response of the microphone was observed to saturate for negative sound pressure values. The normalised synthetic signals were clipped between −0.8 and 1 to emulate this effect. The lack of low-frequency background noise in the synthetic gunshots was compensated for by adding Brownian noise. Equalisation was carried out on the synthetic signals to simulate the resonant properties of the physical gun which also affects how it sounds.

5.3 Methodology and Subjects

A pilot AB test with 10 subjects indicated that the differences between real and synthetic gunshot sounds are clearly discriminable. It was therefore deemed appropriate to assess the perceived realism of the synthetic gunshot sounds in comparison with real recordings.

The listening task involved presenting a hidden reference (i.e. the real recording), a hidden anchor (i.e. a processed version of the reference low-pass filtered at 1.5 kHz) and the stimulus under test (i.e. the synthetic gunshot sound generated as described above).

The subjects' task in each trial was to listen to all stimuli and rate them for their realism using three sliders on a continuous scale from 0 (very unrealistic) to 100 (very realistic). Intermediate markers were also provided: 25 (somewhat unrealistic), 50 (neither realistic nor unrealistic), and 75 (somewhat realistic). The order of different tested gun types as well as the presentation order in each trial was randomised. Subjects were instructed to use the headphones that they typically use while playing games. The collected data are normalised and reported in the range 0–1. Figure 13 shows the test interface employed in the listening test.

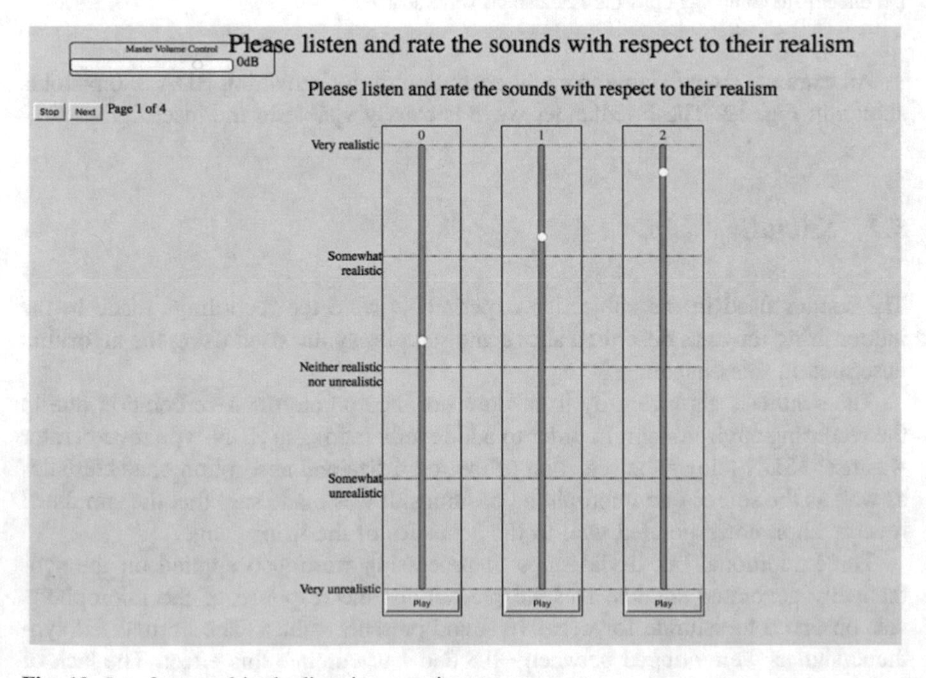

Fig. 13 Interface used in the listening experiment

The evaluation was carried out online using the Web Audio Evaluation Toolbox (Jillings et al. 2015) over listeners employed via a paid experiment crowdsourcing service. A total of 60 subjects (42 male and 18 female) between the ages of 18 and 47 ($M = 27.7$, SD = 6.4) with self-reported normal hearing participated in the test. The online screening process allowed the selection of subjects who regularly play video games. The selected subjects self-reported playing video games 17 h per week on average.

5.4 *Results*

It was noticed during the analysis that one of the subjects reported only two out of the presented four gun types. This subject was removed from the analysis. This meant that there were $N = 59$ dependent samples for each comparison.

The marginal means for the real recordings (REAL), synthetic sounds (SYNTH), and the anchor (ANC) are 0.6372 (SD = 0.2452), 0.4793 (SD = 0.2495), and 0.3428 (SD = 0.2646), respectively. A paired t-test was carried out to compare the means of REAL, SYNTH, and ANC. The test indicated that the differences between the means of the three different cases are statistically significant at $\alpha = 0.05$ level with $t(235) = 7.466$, $p < 0.001$ (REAL-SYNTH), $t(235) = -11.929$, $p < 0.001$ (ANC-REAL), and $t(235) = -5.681$, $p < 0.001$ (ANC-SYNTH). This shows that real recordings were rated higher on average than synthetic sounds, which in turn were rated higher than the anchor.

A second set of analyses were carried out to find out whether this finding holds also for different gun types. The analyses indicated that except for the case of Browning BDA 380 (REAL-SYNTH) and Rossi Magnum R971 (SYNTH-ANC), differences between REAL, SYNTH, and ANC were statistically significant. In the case of Browning BDA 380, the difference between means for REAL ($M = 0.6076$, SD = 0.2648) and SYNTH ($M = 0.5281$, SD = 0.2485) was not significant $t(58) = 1.964$, $p = 0.054$. In the case of Rossi Magnum R971 the difference between means for SYNTH ($M = 0.4019$, SD = 0.2552) and ANC ($M = 0.3490$, SD = 0.2603) was not significant $t(58) = 1.089$, $p = 0.281$.

A third set of analyses was carried out to find out whether synthetic gunshot sounds with a shock wave component (Glock 19c and Rossi Magnum R971) were rated differently than those without (Browning BDA 380 and Glock 21). The results of an independent samples t-test indicate that the mean responses between these groups for were not statistically significantly different at $\alpha = 0.05$ level for both REAL and ANC. However, SYNTH was rated differently and sounds with a shock wave component ($M = 0.4419$, SD = 0.2551) were rated lower than those without ($M = 0.5168$, SD = 0.2390) with $t(234) = 2.328$, $p = 0.021$. The results obtained in the listening test are summarised in Fig. 14.

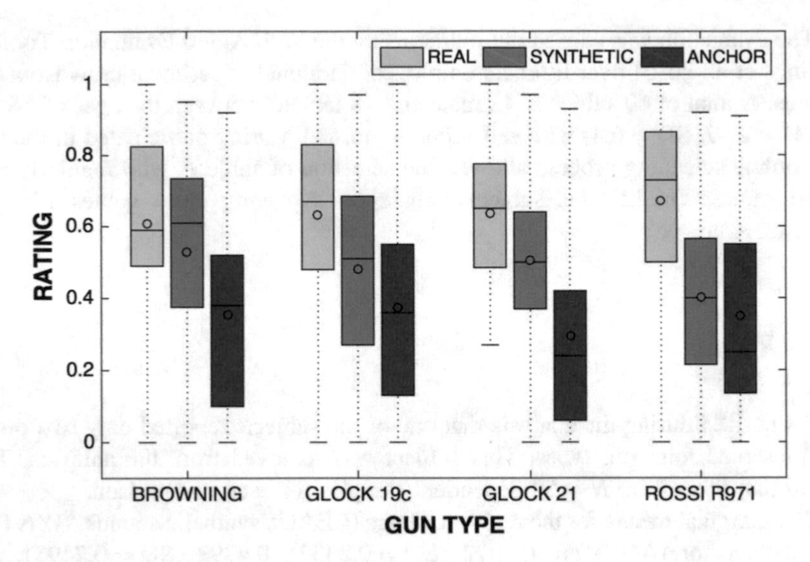

Fig. 14 Boxplots showing the results of the listening test

6 Discussion and Conclusions

The results of the listening test are promising in the sense that while the reference stimuli (REAL) were rated higher than the test stimuli (SYNTH), the difference was not very high. This is an indicator that the described procedural synthesis algorithm can generate usable weapon sounds and can eliminate the need for costly and time-consuming recordings. Another advantage is the ability to separate the source from the environment making it possible to use and reuse content in different virtual environments.

There were some surprising results from listening test: The reference signals were rated on average as being even less than "somewhat realistic". Since these sounds are actual recordings, the expectation was that the average should have been closer to "very realistic". This may indicate that the subjects have different expectations of real gunshot sounds.

The synthesis model has certain limitations, however. Ballistic information is neither readily nor comprehensively available from ammunitions manufacturers. Similarly, different modifications to the guns (such as the Glock 19c in this study which has a compensator for reducing recoil) may make it harder to tune the model to obtain which can match the real recordings exactly. It is also impossible to measure the resonance characteristics of the gun with any accuracy.

Evaluation of procedurally generated gunshot sounds also has some limitations. A real gunshot sound can easily exceed 150 dB SPL at reasonable recording distances. Reproduction of such a sound level over typical consumer-grade headphones or loudspeakers is both technically impossible and ethically unacceptable as this could potentially damage hearing. Another important limitation is the fact that the two components of the gunshot have very short durations and they cannot reliably be tested without adding reverberation that may have a confounding effect on the results.

Despite these limitations, the proposed method is a good starting point for more elaborate algorithms that can also take into account additional components of gunshot sounds such as the mechanical sounds.

Acknowledgements The author wishes to thank MKEK shooting range manager, Mr. Tarık Anıl, as well as indoor shooting range personnel for their help and MKEK for allowing the usage of their facilities for the gunshot recordings used in the subjective experiments. The author also wishes to thank Mr. James D. Johnston for the advice on the recording set-up.

References

BF4 audio files. Symthic Game Science Forums (2014). http://forum.symthic.com/battlefield-4-general-discussion/6591-bf4-audio-files/

Beck, S.D., Nakasone, H., Marr, K.W. (2011). Variations in recorded acoustic gunshot waveforms generated by small firearms. J. *Acoust. Soc. Am.*, 129(4), 1748–1759

Carlucci, D.E., Jacobson, S.S. (2013). *Ballistics: theory and design of guns and ammunition*. CRC Press

Cascone, K., Petkevich, D.T., Scandalis, G.P., Stilson, T.S., Taylor, K.F., Van Duyne, S.A. (2005). *Apparatus and methods for synthesis of internal combustion engine vehicle sounds,* US Patent 6,959,094

De Sena, E., Hacıhabiboğlu, H., Cvetkovic, Z., Smith, J.O. (2015). Efficient synthesis of room acoustics via scattering delay networks. IEEE/ACM Trans. on Audio, Speech, and Language Process. 23(9), 1478–1492

Electronic Arts (2016) Battlefield. https://www.battlefield.com

Fansler, K.S., Thompson, W.P., Carnahan, J.S., Patton, B.J. (1993). A parametric investigation of muzzle blast. Tech. Rep. ARL-TR-227, US Army Research Laboratory, Arlington, VA, USA

Farnell, A. (2010). *Designing sound*. MIT Press

Friedlander, F.G. (1946). The diffraction of sound pulses. I. Diffraction by a semi-infinite plane. *Proc. Royal Soc. London, A: Mathematical, Physical and Engineering Sciences*, 186(1006), 322–344

Gearbox Software (2009). Borderlands. http://borderlandsthegame.com/

Huopaniemi, J., Savioja, L., Karjalainen, M. (1997). Modeling of reflections and air absorption in acoustical spaces: a digital filter design approach. *Proc. IEEE Workshop on Appl. of Signal Process. to Audio and Acoust.* (WASPAA '97)

International Organization for Standardization (1993). Acoustics-Attenuation of sound during propagation outdoors, Part 1: Calculation of the absorption of sound by the atmosphere, ISO 9613-1:1993)

Jot, J.M., Chaigne, A. (1991). Digital delay networks for designing artificial reverberators. *Proc. 90th Convention of the Audio Eng. Soc.*, p. Preprint # 3030. Paris, France

Kahrs, M., Avanzini, F. (2001). Computer synthesis of bird songs and calls. *Proc. Conf. on Digital Audio Effects (DAFx-01)*. Limerick, Ireland

Kuttruff, H. (2009). *Room acoustics*. CRC Press

Maher, R.C. (2009). Audio forensic examination. *IEEE Signal Process*. Mag. 26(2), 84–94

Maher, R.C., Shaw, S.R. (2008). Deciphering gunshot recordings. *Proc. 33rd Int. Conf. of the Audio Eng. Soc.: Audio Forensics-Theory and Practice*, pp. P–2. Audio Engineering Society, Denver, Co, USA

Maher, R.C., Shaw, S.R. (2010). Directional aspects of forensic gunshot recordings. *Proc. 39th Int. Conf. of the Audio Eng. Soc.: Audio Forensics: Practices and Challenges*, 4-2. Audio Engineering Society, Hillerod, Denmark

Nicholas Jillings Brecht De Man, D.M., Reiss, J.D. (2015). Web audio evaluation tool: A browser-based listening test environment. *Proc. 12th Sound and Music Computing Conference*. Maynooth, Ireland

Nordahl, R., Turchet, L., Serafin, S. (2011). Sound synthesis and evaluation of interactive footsteps and environmental sounds rendering for virtual reality applications. *IEEE Trans. Visualization and Computer Graphics,* 17(9), 1234–1244

Rasmussen, P., Flamme, G., Stewart, M., Meinke, D., Lankford, J. (2009). Measuring recreational firearm noise. *Sound and Vibration*, Mag. 43(8), 14–18

Schröder, M. (2009). Expressive speech synthesis: Past, present, and possible futures. J. Tao, T. Tan (eds.) *Affective Information Processing*, pp. 111–126. Springer London

Spratt, K., Abel, J.S. (2008). A digital reverberator modeled after the scattering of acoustic waves by trees in a forest. *Proc. 125th Convention of the Audio Eng. Soc.*, p. Preprint # 7650. Audio Engineering Society, San Francisco, CA, USA

Statista Inc. (2016). Genre breakdown of video game sales in the united states in 2015. https://www.statista.com/statistics/189592/breakdown-of-us-video-game-sales-2009-by-genre/

Stoughton, R. (1997). *Measurements of small-caliber ballistic shock waves in air. J. Acoust. Soc. Am.*, 102(2), 781–787

Team Stack Studios (2015). Stack gun heroes. http://www.stackgunheroes.com

US Department of Defense (2002) America's army https://www.americasarmy.com

Välimäki, V., Parker, J., Savioja, L., Smith, J.O., Abel, J. (2016). More than 50 years of artificial reverberation. *Proc. 60th Int. Conf. of Audio Eng. Soc.*: DREAMS (Dereverberation and Reverberation of Audio, Music, and Speech), pp. K–1. Leuven, Belgium

Välimäki, V., Parker, J.D., Savioja, L., Smith, J.O., Abel, J.S. (2012). Fifty years of artificial reverberation. *IEEE Trans. on Audio, Speech and Language Process.*, 20(5), 1421–1448

Vorländer, M. (2007). *Auralization: Fundamentals of acoustics, modelling, simulation, algorithms and acoustic virtual reality*. Springer Science & Business Media

Whitham, G.B.: The flow pattern of a supersonic projectile (1952). *Comm. on Pure and Appl. Math.*, 5(3), 301–348

Author Biography

Hüseyin Hacıhabiboğlu is an associate professor of signal processing and chair of the Department of Modelling and Simulation at the Graduate School of Informatics in METU, Ankara, Turkey, where he coordinates the graduate program on multimedia informatics and teaches audio signal processing, procedural sound design, and game metrics. He is a senior member of the IEEE, a member of the Audio Engineering Society (AES), and the European Acoustics Association (EAA).

He received the B.Sc. (honors) degree from the Middle East Technical University (METU), Ankara, Turkey, in 2000, the M. Sc. degree from the University of Bristol, Bristol, UK, in 2001, both in electrical and electronic engineering, and the Ph.D. degree in computer science from Queen's University Belfast, Belfast, UK, in 2004. He held research positions at University of Surrey, Guildford, UK (2004–2008) and King's College London, London, UK (2008–2011). His research has always concentrated on applications of signal processing in audio and acoustics. He is named as the co-inventor in three patents on audio source separation, multichannel audio, and artificial reverberation. The latter is being used in a AAA video game title which will be released shortly.

Hacıhabiboğlu started playing games as a kid on his Commodore 64 and has always been fascinated by game audio. His most popular game genres are shooters and cinematic adventure games. His personal interest in games, coupled with his research interests in audio signal processing and acoustics, led to the study reported in the chapter on procedural sound in this volume.

Dynamic Player Pairing: Quantifying the Effects of Competitive Versus Cooperative Attitudes

Gerry Chan, Anthony Whitehead and Avi Parush

Abstract The cooperative or competitive attitudes of players in video games can influence their enjoyment of the game. We examined the effects of matched and unmatched attitude pairings of players on their enjoyment within cooperative and competitive game scenarios. We tested the hypothesis that matching individual attitudes to game scenarios would increase game enjoyment. Sixty-two participants (31 pairs) played virtual Bocce and completed a questionnaire that evaluated their level of enjoyment and their competitive and cooperative attitudes. Quantitative results showed that cooperative pairings enjoyed playing together and male pairings enjoyed playing most in a competitive game scenario. Qualitative results showed that enjoyment was evident when players exchanged signs of encouragement and when they experienced winning. Implications for game design could include initiating a survey or engineering interactions to detect a player's level of competitiveness or cooperativeness in online games to improve game enjoyment.

1 Introduction

Not only game contents can be created procedurally—player matching should be realized with a dynamic method, too. The goal of this work is to explore the possibility of pairing individuals of similar versus dissimilar attitudinal orientations, for increasing enjoyment to ensure persistent engagement in a video game. In particular, we conducted a study testing how one's tendency towards competition or cooperation influences engagement. It also examined how the presence of another

G. Chan (✉) · A. Whitehead · A. Parush
Carleton University, Ottawa, Canada
e-mail: gerry.chan@carleton.ca

A. Whitehead
e-mail: anthony.whitehead@carleton.ca

A. Parush
e-mail: avi.parush@carleton.ca

© Springer International Publishing AG 2017
O. Korn and N. Lee (eds.), *Game Dynamics*,
DOI 10.1007/978-3-319-53088-8_5

individual of similar or dissimilar attitude, and the game type influences engagement.

An important motivator for engagement is the *social* aspect of an activity. Social support from family members and co-exercisers keeps exercise participation rates higher and keeps individuals engaged longer term (Carron et al. 1996). Similar results are found in the context of video games (Marker et al. 2015). Increasing social interactions (Ekman et al. 2012) and benefits (Granic et al. 2014) motivates players to play video games. We see an opportunity for strangers to be brought together dynamically to play videogames in a way that increases the motivation between multiple players.

A second motivator for engagement in games is *enjoyment*. In quantitative studies measuring intrinsic motivation, the concept of game enjoyment is commonly associated with an individual's experience of fun and their level of interest (Mekler et al. 2014). In this study, because of our interest in paired play, enjoyment is conceptualized more inter-relationally as an individual's experience of fun during play and the degree to which an individual enjoyed their interactions with their fellow player. Enjoyment in this context is not determined solely by the game itself, but also by the experience of the game with another. We call this latter factor "social influence." This study therefore examined two key factors for increasing motivation: (1) enjoyment and (2) social influence, by pairing individuals of similar versus dissimilar attitudes. As online games proliferate, there will be a wide variety of possible player pairings, and it is increasingly important to understand how these pairings may or may not encourage long-term participation in a game.

To explore how we can encourage motivation beyond the novelty of a new technology and maintain interest in the game, we hypothesized that matching individuals with similar attitudes while also matching attitudes to game scenarios would increase game enjoyment. For example, we wanted to test whether a competitive individual would enjoy playing with another competitive individual (two individuals with a similar attitude) in a competitive game scenario (a game that matches the attitudes of the individuals). In this situation, both the game scenario and second player act as motivators for the player, keeping them engaged because the player will not only enjoy the game itself, they will also enjoy their interaction with the other player.

In our work, we adopted Deutsch's (2006) approach which treats the competitive—cooperative spectrum as a specific "attitude" among other dimensions.[1] "Attitude", according to Deutsch (2006), refers to the tendency to respond "evaluatively, favorably, or unfavorably, to aspects of one's environment or self" (Deutsch 2006). The basic psychological orientation of cooperation implies a positive attitude that people benefit one another, whereas competition implies a negative attitude that people are against one another (Deutsch 2006). Competition and cooperation are also

[1]Substitutability, attitudes, and inducibility are the three dimensions that Deutsch (2006) identified to describe the competitive–cooperative spectrum. All three are necessary for understanding the social and psychological processes involved in creating the major effects of competition and cooperation.

conceptualized as opposing goal structures that motivate behavior in social situations (Deutsch 1949).

The research question of this study is: would pairing people with similar attitudes increase enjoyment within the context of a video game play that matched the attitudes of the players to the game, as opposed to unmatched pairings and an unmatched game type? A game with low physical requirements was purposefully selected as a test bed for this question because we wanted to eliminate confounding factors related to ability and familiarity. Virtual Bocce created by *Sony Sports Champions* (Sony Entertainment America 2010) offered a game that our participants probably did not know well and playing scenarios congruent to our study: a purely competitive (head-to-head) and purely cooperative (same team) playing options. In addition to attitudes, possible effects of gender and the prior relationship (whether pairs were friends or strangers) between player pairs were also analyzed.

2 Literature Review and Related Works

In this literature review we summarize research to date relating to gaming enjoyment and engagement, especially as it relates to the more psychological aspects of competitive and cooperative attitudes and competitive and cooperative games.

A great deal of effort has been invested by the game industry to understand the psychology of player modeling (Cowley and Charles 2016) and to develop appropriate matchmaking algorithms with the ultimate goal of creating the best possible experience for players in online game contexts (e.g.: Nacke et al. 2014; Farnham et al. 2009; Riegelsberger et al. 2007; Sherry et al. 2006). Of particular interest here, in an active video game context, Mellecker et al. (2013) suggested that validated personality tests should be applied to exercise game play for experiencing enjoyment. But what is enjoyment?

From a psychological perspective, “enjoyment” is typically associated with intrinsic motivation [doing an activity simply for the enjoyment of the activity itself, rather than for external rewards or pressures (Ryan and Deci 2000)] and extrinsic motivation [doing an activity in order to attain some separable outcome or value (Ryan and Deci 2000)] motivation (Przybylski et al. 2010). “Engagement” is also associated with intrinsic and extrinsic motivation (Przybylski et al. 2010), as well as the experience of flow [an intense state of concentration which amounts to absolute absorption of the activity (Csikszentmihalyi 1990; Gregory 2008)]. Both enjoyment and engagement are important concepts that are continuously considered when designing good video games. Enjoyment/engagement and motivation are linked but the relationship can be complex.

Although a competitive activity motivates behavior, it also decreases intrinsic motivation when that activity is perceived as a goal to achieve a reward (Deci et al. 1981). In a competitive exercise game context, intrinsic motivation increases for highly competitive individuals, but decreases for less competitive individuals (Song et al. 2009, 2013). In contrast, a cooperative exercise game results in higher levels of intrinsic motivation for both types of players than a competitive exercise game (Staiano et al. 2012) and induces more commitment between friends compared to strangers (Peng and Hsieh 2012). Via these dynamics related to motivation, the nature of the activity has an influence on enjoyment and engagement.

Social presence also increases video game enjoyment (Gajadhar et al. 2008). For example, competitive multiplayer videogames are enjoyable because they offer a social situation for effective competition (Schmierbach et al. 2012). Cooperative play, while also enjoyable, is not as enjoyable as competitive play (Schmierbach et al. 2012). As with activity type, social presence may influence enjoyment through motivation. Players may be motivated by their engagements with others. In exercise games, we know *social facilitation* "the effect of the presence of others in the individual" (Bond 2001), motivates competitive individuals to perform harder (Snyder et al. 2012) and playing with another person is more enjoyable and motivational than playing alone (Peng and Crouse 2013).

But what is known of gameplay behaviors themselves with respect to competition and cooperation? In social video games, competitive and cooperative play with other players encourage helping behaviors (Velez and Ewoldsen 2013). In a violent, multiplayer video game, cooperative playing scenarios elicited cooperative behaviors (Ewoldsen et al. 2012; Greitemeyer et al. 2012). In third-person, multiplayer video games, cooperative play increases cooperative behaviors, competitive play increases aggressive behaviors, and pre-existing relationships between players does not result in more cooperative behavior (Waddell and Peng 2014).

Prior relationships between the players may also play a role in game appeal. In both collocated and online social games, among friends, a cooperative game is more appealing compared to a competitive game, competitive games are particularly appealing for males relative to females, and the appeal of cooperative games decreases if it is played between strangers as opposed to friends (Embaugh 2016).

Our study builds on these results and examines the effects of pairing players with similar and dissimilar attitudes on the competitive-cooperative continuum on enjoyment in either competitive or cooperative video game scenarios.

3 Evaluation

3.1 Design and Participants

A 2 by 3 within participant, counter-balanced design (Table 1) was used to test our hypotheses. $N = 62$ individuals (31 pairs).

University students were recruited from the Carleton community by posters and email notices sent to professors in various disciplines of study, asking them to publicize the study in their class. Sixty-two participants ranging in age from 18 to 57 years old ($M = 27$ years, $SD = 8$ years) volunteered to participate in the study. To ensure two players in all experimental session, participants were asked to bring a

Table 1 Experimental design: 2 game scenarios, 3 attitude pairings, 62 participants

	Attitude pairings		
Game scenarios	Competitive versus competitive	Cooperative versus cooperative	Competitive versus cooperative
Competitive	7	7	17
Cooperative	7	7	17

Table 2 Gender and attitude pairings of the 31 player pairs

	Attitude pairings			
Gender pairings	Competitive versus competitive	Cooperative versus cooperative	Competitive versus cooperative	Total
Male versus male	1	2	6	9
Female versus female	2	4	2	8
Male versus female	4	1	9	14
Total	7	7	17	31

friend along. Those who were not able to bring a friend were paired with another participant who was a stranger to them. Fifty-one participants had not played the game before and twenty-five participant pairs were friends.[2] On a 10-point scale (1 = novice to 10 = expert), there was a mixture of novices and experts on game experience ($M = 5.1$, $SD = 2.7$). The numbers of gender pairings with respect to different attitude pairings (Table 2) were not balanced because participants volunteered to participate on a random basis, and individual attitudes and pairings were formulated after preliminary data collection using a median split. There was no significant relationship between gender and attitude pairings, $\chi^2(4) = 7.59$, $p = 0.11$.

3.2 Measures

3.2.1 Enjoyment and Social Influence

A post-game questionnaire, using a 7-point Likert scale (1 = strongly disagree to 7 = strongly agree) measured play enjoyment and social influence. Custom items (Table 3) were specifically developed to evaluate enjoyment and social influence

[2]Friends and non-friends could influence the results due to friendship biases. For example, friends could be more supportive or engage in more conversation with each other during play compared to non-friends.

Table 3 Post-game survey statements that evaluated level of enjoyment and social influence

Enjoyment	Social influence
I liked playing on the same team	I felt pressured to win because of the other person
I liked playing head-to-head	I felt that the other person was a good competitor
I thought that playing with another person was fun	I felt that my partner was helpful when we were playing together against the computer
I enjoyed playing with another person	I was motivated to play better because of the other person
I enjoyed playing against another person	I would have rather played alone

because we were interested in exploring: game scenario, social influence, motivation, game type, and personal preference, all of which could influence the playing experience.

In addition to a post-game questionnaire, play enjoyment was captured qualitatively. The researcher took on the role of "complete observer" (Gold 1958) and employed "semi-structured observations"—of controlled simulations or analogue situations (Ostrov and Hart 2013) to collect indicators of enjoyment. Collecting such indicators is difficult. Even tape/video recorders are unable to capture all the relevant aspects of social processes (Babbie and Benaquisto 2002). To consider a broad range of factors, we designed a standardized observation and interpretation record sheet to capture and organize any relevant signs of enjoyment. Both supportive (e.g. pat on back or firm hand shake) and unsupportive (e.g. avoidance) gestures, together with facial expressions (e.g. smile) were collected as indicators of enjoyment. In addition to gestures, dialogues between participant pairs were transcribed verbatim for the purpose of analysis. Tone of voice and the words chosen to express one's thoughts or feelings about the playing experience were analyzed.

3.2.2 Competitive/Cooperative Attitudes

Individual attitudes for use of competitive (e.g. to succeed, one must compete against others) versus cooperative (e.g. I enjoy working with others to achieve joint success) strategies for success were evaluated with a questionnaire using a 5-point Likert scale (1 = always to 5 = never) developed by Simmons et al. (1988). Higher scores indicate lower levels of competitiveness or cooperativeness as an attitude. All 24 items are unbiased relative to sex or age and the entire scale yields reasonable statistical reliability ($r = 0.78$). The questionnaire was integrated into a post-game questionnaire and was administered after participants completed the game.

3.3 Apparatus

Virtual Bocce was projected onto a white wall using a 1080p multimedia projector. The projector was placed on top of a mobile cart and was connected to a Sony PlayStation 3 to run the game. A set of multimedia speakers were connected to the game console for better sound. An eye camera tracked the movements of the player and a red "X" was taped to the floor to remind participants where to stand during the experiment for best motion sensing results. Participants held on to a Move motion controller to interact with virtual objects in the game. All play sessions were recorded using a video camcorder mounted on top of a tripod. Figure 1 illustrates the location of each apparatus, some approximate measurements, and the participant's position during the experiment. Figure 2 shows a pair of players in a study session.

3.4 Procedures

After receiving clearance from the university ethics committee, the recruitment for participation began. People who contacted the researcher and expressed interest in participating were scheduled for a test session. The experiment was 1 h and was divided into three parts: a practice session followed by two experimental sessions.

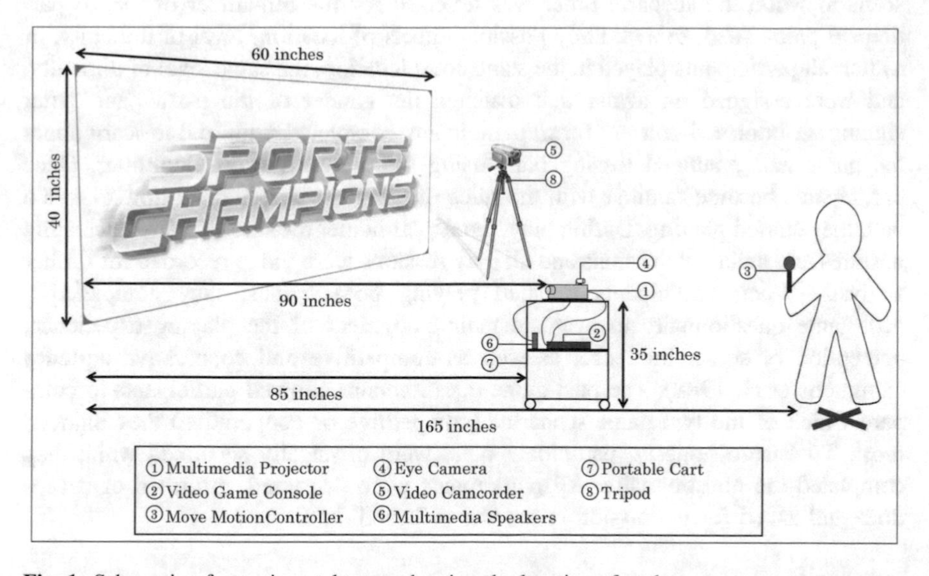

Fig. 1 Schematic of experimental setup showing the location of each apparatus, measurements, and position of participant

Fig. 2 Photo of player pairs playing virtual Bocce during a study session (faces of players have been purposefully blurred to protect their identity)

It was presented to participants as a usability test to capture their natural behaviors as much as possible. All participants played the game in pairs. To avoid order effects, 15 participant pairs played a competitive scenario followed by a cooperative scenario while the scenario order was reversed for the remainder of the 16 participant pairs. Also, to avoid any possible effects of location, level of difficulty, or avatar, all participants played in the same court location, the same level of difficulty, and were assigned an avatar that matched the gender of the participant. After signing an informed consent form, participants completed a tutorial to learn about the game and practiced tossing balls using the Move Motion Controller. Once participants became familiar with the rules of the game and learned how to toss a ball, they started playing. During play, the experimenter took note of any interesting gestures and facial expressions and all play sessions were video recorded for further analysis. When participants finished playing both games, they completed a post-game questionnaire to evaluate their enjoyment of the playing experience, perception of social influence, as well as competitive and cooperative attitudes (Simmons et al. 1988). The post-game questionnaire allowed participants to compare which of the two game scenarios (competitive or cooperative) they enjoyed most. To ensure honesty, participant pairs were physically separated while they completed the questionnaire. All participants were debriefed, informed of deceptions and asked for permission to use the collected data.

4 Results

Data were computed using IBM SPSS Version 22.0 and prior to computation of statistics, data were screened for missing values, outliers and out-of-range values. Subjective results were derived from the analysis of the post-game questionnaire. Non-parametric tests were selected to analyze effects of individual attitudes, attitude pairings, gender pairings and relationship because an ordinal scale was used in the post-game questionnaire and the distribution of data was skewed (Table 4). Figure 3 shows the descriptive statistics for overall enjoyment ($M = 5.87$, $SD = 1.21$) and social influence ($M = 4.37$, $SD = 2.00$). The alpha was set at 0.05 for all statistical tests.[3]

Qualitative results were derived from the analysis of playing sessions and a further review of video recordings. A median split was computed on the competitive/cooperative preferences scale (Simmons et al. 1988) to classify individual attitudes with regard to their proclivity to favor the use of competitive/cooperative strategies. A median split is a method for converting a continuous variable into a categorical variable and is commonly used in psychology to enable the analysis of group differences rather than individual differences (MacCallum et al. 2002), as is the case here. Drawbacks associated with the use of a median split include: loss of information regarding individual differences and lower effect size, power, and measurement reliability (MacCallum et al. 2002). In the end, we had classified thirty-one individuals as individuals who favored the use of competitive strategies and thirty-one who favored cooperative strategies. Attitude pairings (competitive match = two competitive individuals; cooperative match = two cooperative individuals; mismatch = one competitive individual and one cooperative individual) were then generated by comparing the preference scores of two individual participants. Seven competitive matched pairings, seven cooperative matched pairings, and seventeen mismatched pairings were formed (Table 1).

4.1 *Effect of Individual Attitudes*

To investigate if there was an effect of individual attitudes with respect to game scenario, a median test was conducted on questions in the post-game questionnaire that evaluated game scenario enjoyment. Results (Table 5) showed that both competitive and cooperative individuals did not enjoy playing in a cooperative game scenario (*Median* = 6.00). Further, competitive individuals did not enjoy playing with another player, as much as, cooperative individuals (*Median* = 6.00).

[3] $^{*}p < 0.05$; $^{**}p < 0.01$.

Table 4 Descriptive statistics for level of agreement (7-point Likert scale: 1 = strongly disagree to 7 = strongly agree) with respect to post-game survey statements

Measure	Total sample size ($N = 62$)	Descriptive statistics				
	Post game survey statements	Mean	SD	Median	Skewness	Kurtosis
Enjoyment	I liked playing on the same team	5.47	1.33	6.00	−1.45	2.11
	I liked playing head-to-head	5.56	1.35	6.00	−1.30	1.03
	I thought that playing with another person was fun	6.37	0.71	6.00	−1.83	7.08
	I enjoyed playing with another person	6.34	0.75	6.00	−1.63	5.22
	I enjoyed playing against another person	5.63	1.42	6.00	−1.27	0.99
Social influence	I felt pressured to win because of the other person	3.50	1.92	3.00	0.36	−1.21
	I felt that the other person was a good competitor	5.69	1.11	6.00	−1.66	3.20
	I felt that my partner was helpful when we were playing together against the computer	5.50	1.24	6.00	−0.88	0.60
	I was motivated to play better because of the other person	5.27	1.33	6.00	−1.04	0.88
	I would have rather played alone	1.89	1.04	2.00	2.21	8.40

4.2 Effect of Attitude Pairings

To examine the effects of attitude pairings on game enjoyment, an intra-class correlation (ICC) [a measure of reliability between scores provided by two or more raters (Cicchetti and Sparrow 1981)] was conducted on questions in the post-game questionnaire on matched, competitive vs. competitive ($N = 14$) and cooperative versus cooperative ($N = 14$), and mismatched, competitive vs. cooperative ($N = 34$), attitude pairings. Results showed that both competitive and cooperative matched pairs enjoyed playing in a competitive game scenario, ICC = 0.843. Results also showed that regardless of whether pairings were matched or mismatched, participants enjoyed playing with another individual, ICC = 0.678. To further examine the effects of attitude pairings on game enjoyment, a

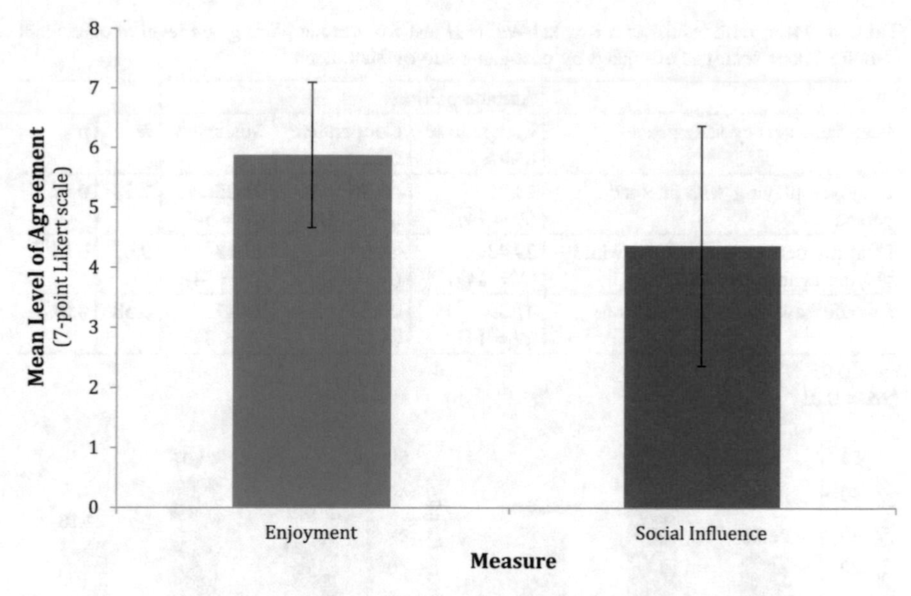

Fig. 3 Means and standard deviations for overall enjoyment and social influence

Table 5 Median test results for individual competitive and cooperative attitudes by post-game survey statements

Overall median = 6.00		Post-game survey statements		
Individual attitudes	Range	I liked playing on the same team	I enjoyed playing with another player	Playing with another player was fun
Competitive	Less than or equal to overall median	29	21	21
	Greater than overall median	2	10	10
Cooperative	Less than or equal to overall median	23	13	13
	Greater than overall median	8	18	18
	x^2	4.29	4.17	4.17
	p	0.04*	0.04*	0.04*

Number of observations are based on level of agreement (7-point Likert scale) with respect to individual attitudes and range
*α = 0.05

Kruskal-Wallis H test was conducted on matched and mismatched pairings with respect to questions in the post-game questionnaire. Results (Table 6; Fig. 4a, b) showed that cooperative match pairings particularly enjoyed playing together ($H = 7.38$, $p = 0.02$) and felt that their partner was helpful when both of them were

Table 6 Mean rank results of a Kruskal-Wallis H test for attitude pairings on level of agreement (7-point Likert scale) as evaluated by post-game survey statements

	Attitude pairings				
Post-game survey statements	Competitive match	Cooperative match	Mismatch	H	p
I enjoyed playing with another person	23.25 (N = 14)	40.36 (N = 14)	31.25 (N = 34)	7.38	0.02*
I felt my partner was helpful while playing against the computer	29.43 (N = 14)	44.07 (N = 14)	27.18 (N = 34)	9.65	0.01**
I would have rather played alone	41.39 (N = 14)	26.54 (N = 14)	29.47 (N = 34)	6.58	0.04*

*α = 0.05
**α = 0.01

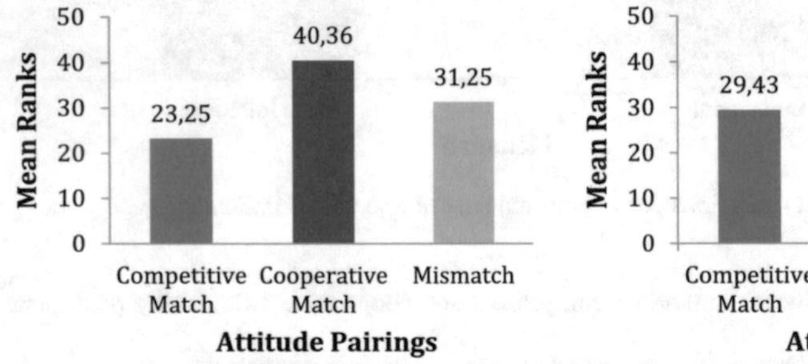

(**a**) I enjoyed playing with another person.

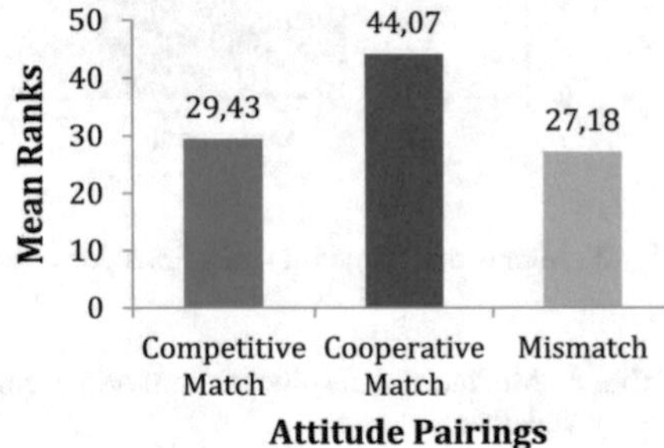

(**b**) I felt my partner was helpful while we were playing against the computer.

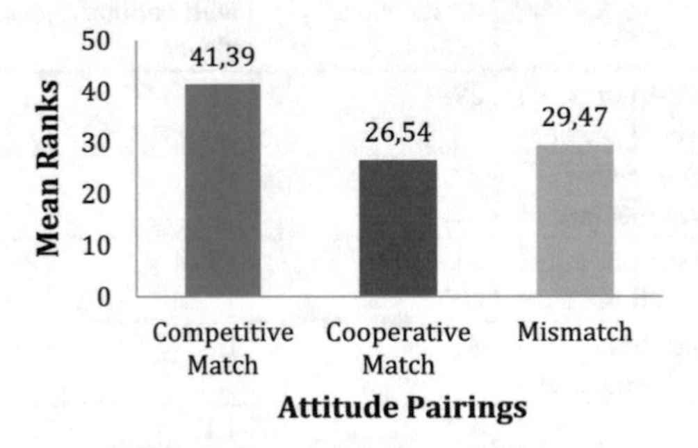

(**c**) I would have rather played alone.

Fig. 4 Mean rank results of a Kruskal-Wallis H test for attitude pairings on level of agreement (7-point Likert scale) as evaluated by post-game survey statements

playing against the computer (H = 9.65, p = 0.01). Results (Table 6; Fig. 4c) also showed that competitive match pairs would have preferred to play alone (H = 6.58, p = 0.04).

From Table 6 and Fig. 4, cooperative individuals enjoyed playing together and felt that their partner was helpful in a cooperative game scenario compared to competitive and mismatched attitude pairings. Furthermore, competitive pairs indicated a preference for playing on their own compared to cooperative and mismatched pairings, suggesting that competitive individuals dislike playing with a second player. This finding supports the theory that competitive individuals focus on individual gain (Deutsch 2006), as the presence of a second player could have: (1) been an impediment to their performance, or (2) required the competitive individual to share resources; which may explain why competitive types would rather play alone.

4.3 Effect of Gender Pairings

To examine if there were any effects of gender pairings (male vs. male, female vs. female, and male vs. female), a Kruskal-Wallis H test was conducted on questions in the post-game questionnaire and results (Table 7; Fig. 5) showed that male versus male ($N = 18$) and male versus female ($N = 28$) pairings most enjoyed playing in a competitive game scenario whereas female vs. female ($N = 16$) pairings least enjoyed playing in a competitive game scenario ($H = 10.67$, $p = 0.01$). Although not statistically significant, mean ranks results (Table 7; Fig. 5) showed that female versus female pairings enjoyed playing in a cooperative game scenario most compared to the other two gender pairing categories ($H = 4.78$, $p = 0.09$).

4.4 Effect of Relationship

To determine if there was an effect of relationship, friends compared to strangers, a Mann-Whitney U test was conducted on a particular question in the post-game

Table 7 Kruskal-Wallis H test results for gender pairings evaluated by post-game survey statements

Post-game survey statements	Gender pairings			H	p
	Male versus male	Female versus female	Male versus female		
Competitive scenario					
I liked playing head-to-head	36.22 ($N = 18$)	19.66 ($N = 16$)	35.23 ($N = 28$)	10.67	0.01**
I enjoyed playing against another player	37.94 ($N = 18$)	21.06 ($N = 16$)	33.32 ($N = 28$)	8.69	0.01**
Cooperative scenario					
I liked playing on the same team	29.86 ($N = 18$)	39.38 ($N = 16$)	28.05 ($N = 28$)	4.78	0.09

Mean ranks show level of agreement on a 7-point Likert scale
**$\alpha = 0.01$

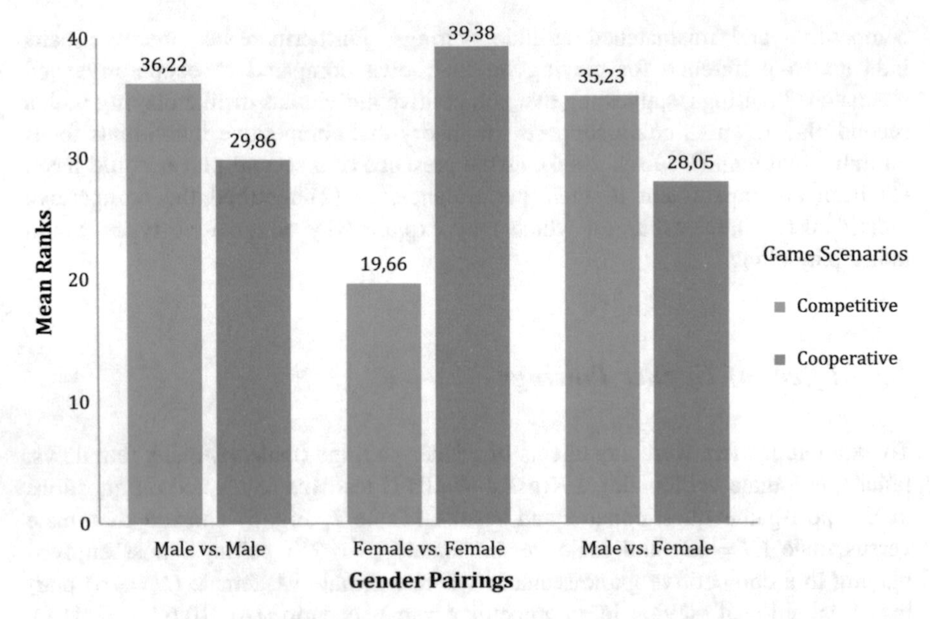

Fig. 5 Mean rank results of a Kruskal-Wallis H test for gender pairings with respect to competitive and cooperative game scenarios

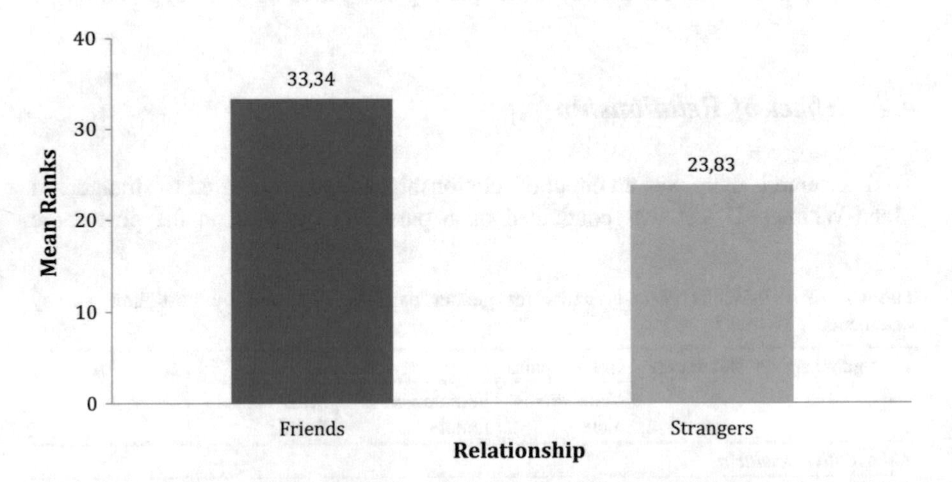

Fig. 6 Mean rank results of a Mann-Whitney U test of relationship between friends and strangers

questionnaire that evaluated whether or not the participant would like to play with the same partner again. Marginally significant results (Fig. 6) showed that friends ($N = 50$) reported they would like to play with the same partner again compared to strangers ($N = 12$), ($U = 208$, $p = 0.07$).

4.5 *Qualitative Analysis*

A thematic analysis approach (Braun and Clarke 2006) was used to identify signs of enjoyment during play. The process of thematic analysis involves working from the data to develop codes, applying these codes across the data set on a line-by-line basis, and then working with these codes to develop broader themes that sufficiently represent the meanings of several codes. Similar to variability in statistics, the codes within a theme reflect variability within that theme. The goal of thematic analysis is ensuring that themes sufficiently represent the range of meanings in a data set. As a result, the major concern in thematic analysis is richness of data rather than amount of data. The use of researcher notes, footage of sessions, and transcripts of interactions in this study resulted in a rich qualitative data set and allowed for triangulation across different data sources in developing themes.

Full playing sessions ranged from 10 to 30 min, and should have been sufficient for participants to experience the game fully. Before generating initial codes, dialogues between participant pairs were transcribed verbatim and initial ideas and notes were recorded by the researcher while developing familiarity with the data. All dialogues were then coded and, exploring these codes across the data set, potential themes were developed. Two major themes emerged as an indication of enjoyment: (1) *positive reciprocation*, and (2) *winning reactions*. "Positive reciprocation" is the exchange of praise between participant pairs when they were performing well in the game (e.g. an exclamation of "good one", when a co-player makes a good toss) and the exchange of support between participant pairs when they were not performing well (e.g. consoling words such as "better luck next time", when a co-player makes a bad toss). "Winning reactions" occurred when a participant won or achieved a good score and expressed enthusiasm, typically indicated by lifting both arms upwards and laughter. The boundaries between both themes are relatively fluid in that positive reciprocation could occur when one player demonstrated a winning reaction upon another player winning, with the other player reciprocating by doing the same in return.

4.5.1 Theme 1: Positive Reciprocation

Positive reciprocation was particularly obvious when participants were discussing strategies for winning the game. The following transcript snippet coded as *positive reciprocation* was taken from a cooperative game scenario where 2 males who were strangers and where one was cooperative and the other competitive conversed about strategies to win the game:

Player A (cooperative): "Yeah, I think we can knock that one out."
Player B (competitive): "Ah, come on, what is this?" [Annoyed at the avatars].
Player A: "Try a little bowling technique, oh, no!" [Laughed] "Worst case scenario."

Player B: "This cannot work for us. We're going to have to try a different strategy; try not to get it away."
Player A: "Good,"
Player B: "Did I get it close enough?"
Player A: "You got it away from theirs."
Player B: "I did, nice!"
Player A: "We get 2 points for that!" [Smiled] "I prefer bowling now, it's more natural!"

In another cooperative game scenario, also coded as *positive reciprocation*, 2 males, who were friends and who both favored the use of competitive strategies for success, shared the following dialogue:

Player A (competitive) made a good shot and player B said: "nice."
Player B (competitive): "I think we're going to win this one."
Player A: "Ok, let's see if we can do this. Not happening; I think we're going to win this one. Ok, they just showed up." When it was player B's turn, player A said: "player B's name [in a serious and supportive way] do this." Player A made a good shot, Player B: "nice."
Player A: "Yeah! That's it, game's done!"
Player B: "Multi-point chance, let's do it…ooo." [Excited, made a good shot].
Player A said: "nice." Player B made a good shot, player A said: "nice, nice, very nice; we are back in the lead." [The avatars were not doing well, both players A and B laughed at them].
Player B made a good shot, player A said: "nice, nice" and player B said: "yes."
Player A laughed when he made a good shot; then, it was player B's turn and said: "go for it." Player B: "I think we got this."

4.5.2 Theme 2: Winning Reactions

Enjoyment was evident when participants laughed, jumped, threw both arms upwards or shook hands. For example, in one competitive scenario, 2 males who were friends (one competitive and one cooperative) behaved this way:

Player A (competitive) made a good shot, laughed and said "yes."
Player A asked player B (cooperative): "Should I spin this one?"
Player B: "Do whatever you want." [Player A went ahead with a spin shot].
Player B: "Match point, you're going to win. See, now I get to throw. Hey, look at that." [Made a good toss].
Player B made another good shot, and Player A held his hand out and wanted to shake hands but, Player B said: "No, no, there's not shaking hands while I'm down 5; I'm going to beat you; never lost a Bocce in PS3 before." Player A made a good shot, said: "yes!" [Laughed, and raised both his arms.]

Player B said "ah, man." Player B made a good shot, player A said: "that was a good shot."

In a different cooperative scenario, 2 females who were friends (one competitive and the other cooperative), used laughter and a high-five to indicate their enjoyment:

Player A (competitive) made a good shot, Player B (cooperative) said: "very nice." Player A made a good shot, both laughed and player B said: "yes!" Player B was relieved that she made a good shot and Player A said: "good." Player B to A: "You can do it." Player B made a good throw and said: "yes!" [Both won at the end, player B initiated a high-five and both laughed!]

From a psychological perspective, the outcome of winning or losing could have elicited an intrinsic reward that would subsequently encourage further play so one may continue to experience the joy of winning. In contrast, positive reciprocation could have extrinsically motivated continuous play because players may have influenced each other due to social facilitation effects.

5 Discussion and Conclusions

Recall that the goal of this study was to explore the effects of competitive and cooperative attitudes on enjoyment in a videogame context. We found evidence that game enjoyment can be enhanced by considering individual attitudes, as well as, other factors such as gender pairings, relationship between gamers, and the nature of the social interactions between players. Subjective results suggest that despite individual attitudes, competitive or cooperative, individual participants did not enjoy playing the cooperative game scenario. Results from attitude and gender pairings suggest that cooperative pairings enjoyed playing together most, whereas competitive pairings would have preferred to play alone, and male pairings enjoyed playing in a competitive scenario most. Additionally, friends enjoyed playing together more than strangers. From qualitative observations, enjoyment was particularly visible when participants learned they had won and when they exchanged signs of encouragement.

Taken together, our results show that player attitudes are important aspects to consider when designing an enjoyable game experience. We found evidence that highly competitive individuals enjoyed playing in a competitive game context while cooperative individuals enjoyed playing together. We have evidence to support the assertion that there is value in assessing player attitudes with respect to their tendency to be cooperative or competitive. For example, in a multiplayer game context, this information can be used to create better player pairings and also to tailor game scenarios to the player types. This will increase (1) the enjoyment of the game itself, and (2) the enjoyment experienced between the players.

The results of this study are in agreement with previous research exploring the importance of social competition for experiencing enjoyment in video games (Schmierbach et al. 2012; Vorderer et al. 2003), as well as player achievement for experiencing enjoyment in exergames (Limperos and Schmierbach 2016). Furthermore, The results extends research concerning the role of competition for increasing motivation in active videogame contexts (Song et al. 2009, 2013). In particular, game scenarios should be tailored to individual differences because—competitive individuals enjoy playing a competitive game, whereas cooperative individuals enjoy playing a cooperative game with other cooperative individuals. With respect to competitive and cooperative attitudes (Deutsch 1949), the finding that competitive pairings preferred to play alone suggests that competitive individuals favor individual success rather than achievement through joint effort. We hypothesized the combination of two competitive individuals could have created a situation where one or both were not willing to share their skills or resources because there was a focus on individual gain (Deutsch 1949). This study also found evidence that cooperative individuals enjoyed playing together in a cooperative game context. This finding is consistent with Deutsch's (1949) theory that cooperative individuals hold positive attitudes in situations where people benefit one another. From a videogame design perspective, the cooperative experience could have increased the interaction between player pairs.

5.1 *Game Design Implications*

Game designers (Table 8) could involve embedding a survey to detect individual attitudes of players into a massively multiplayer online role-playing game such as

Table 8 Recommendations for game design

Results	Design recommendations
Effect of individual attitudes	Introduce a second player for one who favors the use of cooperative strategies for success
	Offer a competitive scenario for one who favors the use of competitive strategies for success
Effect of attitude pairings	Offer a cooperative scenario if two players favor the use of cooperative strategies for success
	Offer individual or head-to-head competitive scenarios if two players favors the use of competitive strategies for success
Effect of gender pairings	Offer a competitive scenario for male versus male and male versus female pairings
	Offer a cooperative scenario for female versus female pairings
Effect of relationship	Offer any game scenario for friends
Qualitative results	Offer opportunities for social interaction
	Offer rewards in which one can celebrate when one wins

"Ultima Forever: Quest for the Avatar" (2013) and then use the results throughout the game, especially where the system automatically tailors scenarios. When setting up games, the gaming systems could recommend other players who possess similar attitudes to increase enjoyment for both. The results of the present study address the "psychosocial research agenda" (Mellecker et al. 2013) by finding evidence that evaluating a player's level of competitiveness and cooperativeness is valuable for creating an enjoyable game experience. Moreover, to increase social interaction between players, perhaps exergames could offer play scenarios that require players to share information, as well as encourage conversation and prosocial behaviors to achieve high scores and to win the game.

Player modeling (Cowley and Charles 2016) and preferences (Sherry et al. 2006) could also be used to determine competitive and cooperative attitudes over time thereby allowing better player pairings in the future. Post-game questions (e.g.: Did you enjoy playing with another player?) might help determine the attitude of a new player when paired with existing players of known attitudinal orientation. Embedded activities in a single player training level can also help determine if a player is competitive or cooperative through interaction with non-player characters (NPCs) who are purposefully designed to effectively elicit competitive or cooperative responses.

Regardless of how overtly or covertly a game algorithm assesses the competitiveness or cooperativeness of a player, knowing this information helps in pairing players together for online play that is more likely to be enjoyable and keeps players engaged and therefore coming back for longer periods of time.

5.2 Limitations and Future Work

One limitation of this study is the disproportion of attitude pairing combinations. To improve the balance between pairings, a future study could aim to recruit 90 participant pairs divided into 3 groups (e.g.: 30 competitive matches, 30 cooperative matches, and 30 mismatches) for a better comparison between each pairing category. A future study could address the issue of gender imbalance by also obtaining a balance here. In addition to the imbalance of attitude and gender pairings, the type of relationship between gamers (friends or strangers) was also unequal—there were more pairs of friends ($N = 25$) compared to pairs of strangers ($N = 6$). Thus, friendship biases could have influenced the enjoyment ratings despite attitude pairings and game scenario matching. Furthermore, this study also considered subjective experiences through the use of qualitative techniques. Future research could include additional objective measure such as physiological measures (e.g. heart rate or galvanic skin response) because they provide good validation for subjective reports in a co-located, collaborative play context (Mandryk and Inkpen 2004). A limitation in the qualitative analysis was that only one researcher coded the data. Additional coders would have enhanced the credibility of this research. Finally, this study tested the plausibility of attitude pairings using a game with relatively low-physical requirements, thus, another avenue for future research could examine the role of competitive and cooperative pairings using a more physically demanding exercise game.

Overall, there could be other individual attitudes and pairing possibilities with game types that remain unexplored. For example, from a personality perspective, a shy, introverted individual may prefer to play alone whereas an outgoing, extroverted individual might enjoy playing with another extrovert. Additionally, a future study may examine the effects of cultural characteristics on paired play. Those who come from a collectivistic culture may enjoy playing cooperative social video games while those who come from an individualistic background may rather play socially competitive games which focus on individual performance and gain.

In closing, this study offers evidence indicating the importance of competitiveness and cooperativeness – both in individual attitudes and the nature of the game—in determining player enjoyment and player influence in videogames. This study provided insight, but more research is needed to uncover other individual and social factors that will increase game enjoyment and engagement.

References

Babbie, E., & Benaquisto, L. (2002). *Fundamentals of Social Research*. Toronto, ON: Nelson Education Ltd.

Bond Jr., C. F. (2001). Social facilitation, psychology of. *International Encyclopedia of the Social & Behavioral Sciences*, 14290–14293.

Braun, V., & Clarke, V. (2006). Using thematic analysis in psychology. *Qualitative Research in Psychology, 3*(2), 77–101.

Carron, A. V., Hausenblas, H. A., & Mack, D. (1996). Social influence and exercise: A meta-analysis. *Journal of Sport and Exercise Psychology, 18*, 1–16.

Cicchetti, D. V., & Sparrow, S. S. (1981). Developing criteria for establishing interrater reliability of specific items: Applications to assessment of adaptive behavior. *American Journal of Mental Deficiency, 86,* 127–137.

Cowley, B., & Charles, D. (2016). Behavlets: A model for practical player modeling using psychology-based player traits and domain specific features. *User Modeling and User-Adaptive Interaction, 26*(2), 257–306.

Csikszentmihalyi, M. (1990). *Flow: the psychology of optimal experience*. New York: Harper and Row.

Deci, E. L., Betley, G., Kahle, J., Abrams, L., & Porac, J. (1981). When trying to win: Competition and intrinsic motivation. *Personality and Social Psychology Bulletin, 7*(1), 79–83.

Deutsch, M. (1949). A theory of cooperation and competition. *Human Relations, 2*, 129–152.

Deutsch, M. (2006). Cooperation and competition. In M. Deutsch, P. T. Coleman, & E. C. Marcus (Eds.), *The Handbook of Conflict Resolution: Theory and practice* (p. 23–42). San Francisco: Jossey-Bass.

Ekman, I., Chanel, G., Järvelä, S., et al. (2012). Social interaction in games: Measuring physiological linkage and social presence. *Simulation & Gaming, 43*(3), 321–338.

Embaugh, K. (2016). Local Co-Op is the most consistently appealing mode of social gaming across gender and age. *Quantic Foundry*. Retrieved from http://quanticfoundry.com/2016/07/21/social-gaming/.

Ewoldsen, D. R., Eno, C. A., Okdie, B. M., Velez, J. A., Gaudagno, R. E., & DeCoster, J. (2012). Effect of playing violent games cooperatively or competitively on subsequent cooperative behavior. *Cyberpsychology, Behavior, and Social Networking, 15*(5), 277–280.

Farnham, S. D., Phillips, B. C., Tiernan, S. L., Steury, K., Fulton, W. B., & Riegelsberger, J. (2009). Method for online game matchmaking using play style information. *U.S. Patent No. 7,614,955*. Washington, DC: U.S. Patent and Trademark Office.

Gajadhar, B.J., de Kort, Y.A.W., & IJsselsteijn, W.A. (2008). Shared fun is doubled fun: Player enjoyment as a function of social setting. In P. Markopoulos, et al. (Eds.), Fun and Games (p. 106–117). *Springer-Verlag Berlin Heidelberg*.

Gold, R. L. (1958). Roles in sociological field observation. *Social Forces, 36*(3), 217–223.

Granic, I., Lobel, A., & Engels, R. C. M. E. (2014). The benefits of playing video games. *American Psychologist, 69*(1), 66–78.

Gregory, E. M. (2008). Understanding video gaming's engagement: Flow and its application to interactive media. *Media Psychology Review, 1*(1).

Greitemeyer, T., Traut-Mattausch, E., & Osswald, S. (2012). How to ameliorate negative effects of violent video games on cooperation: Play it cooperatively in a team. *Computers in Human Behavior, 28*(4), 1465–1470.

Limperos, A. M., & Schmierbach, M. (2016). Understanding the relationship between exergame play experiences, enjoyment, and intentions for continued play. *Games for Health, 5*(2), 100–107.

MacCallum, R. C., Zhang, S., Preacher, K. J., & Rucker, D. D. (2002). On the practice of dichotomization of quantitative variables. *Psychological Methods, 7*(1), 19–40.

Mandryk, R.L., & Inkpen, K. (2004). Physiological indicators for the evaluation of co-located collaborative play. In *Proc. CSCW 2004, ACM Press*, 102–111.

Marker, A. M., & Staiano, A. E. (2015). Better together: Outcomes of cooperation versus competition in social exergaming. *Games for Health Journal*, 4(1), 25–30. doi: 10.1089/g4h.2014.0066.

Mekler, E. D., Bopp, J. A., Tuch, A. N., & Opwis, K. (2014). Systematic review of quantitative studies on the enjoyment of Digital Games. *Session: Understanding and Designing Games, CHI 2014, One of a CHInd, Toronto, ON, Canada*, 927–936.

Mellecker, R., Lyons, E. J., & Baranowski, T. (2013). Disentangling fun and enjoyment in exergames using an expanded design, play, experience framework: A narrative review. *Games for Health Journal: Research, Development, and Clinical Applications, 2*(3), 142–149.

Mythic: Electronic Arts Inc. (2013). *Ultima Forever: Quest for the Avatar*. Retrieved from: http://www.ultimaforever.com/

Nacke, L. E., Bateman, C., & Mandryk, R. L. (2014). BrainHex: A neurobiological gamer typology survey. *Entertainment* computing, 5(1), 55–62.

Ostrov, J. M., & Hart, E. J. (2013). Observational methods. In T. D. Little (Eds.), *The Oxford handbook of quantitative methods in psychology* (p. 285–303). Don Mills, ON: Oxford University Press.

Peng, W., & Crouse, J. (2013). Playing in parallel: The effects of multiplayer modes in active video game on motivation and physical exertion. *Cyberpsychology, Behavior, and Social Networking, 16*(6), 423–427.

Peng, W., & Hsieh, G. (2012). The influence of competition, cooperation, and player relationship in a motor performance centered computer game. *Computers in Human Behavior, 28*, 2100–2106.

Przybylski, A. K., Rigby, C. S., & Ryan, R. M. (2010). A motivational model of video game engagement. Review of General Psychology, 14(2), 154–166.

Riegelsberger, J., Counts S., Farnham, S. D., & Philips, B. C. (2007). Personality Matters: Incorporating Detailed User Attributes and Preferences into the Matchmaking Process. *System Sciences, 2007. HICSS 2007. 40th Annual Hawaii International Conference on, Waikoloa, HI*, 87–87.

Ryan, R. M., & Deci, E. L. (2000). Intrinsic and extrinsic motivations: Classic definitions and new directions. *Contemporary Educational Psychology, 25*, 54–67.

Schmierbach, M., Xu, Q., Oeldorf-Hirsch, A., & Dardis, F. E. (2012). Electronic friend or virtual foe: Exploring the role of competitive and cooperative multiplayer video game modes in fostering enjoyment. *Media Psychology, 15*, 356–371.

Sherry, J. L., Lucas, K., Greenberg, B., & Lachlan, K. (2006). Video game uses and gratifications as predictors of use and game preference. In P. Vorderer & J. Bryant (Eds.), *Playing computer games: Motives, responses, and consequences* (p. 213–224). Mahwah, NJ: Lawrence Erlbaum.

Simmons, C. H., Wehner, E. A., Tucker, S. S., & King, C. S. (1988). The cooperative/competitive strategy scale: A measure of motivation to use cooperative or competitive strategies for success. *The Journal of Social Psychology, 128*, 199–205.

Snyder, A. L., Anderson-Hanley, C., & Arciero, P. J. (2012). Virtual and live social facilitation while exergaming: Competitiveness moderate exercise intensity. *Journal of Sports & Exercise Psychology, 34*, 252–259.

Song, H., Kim, J., Tenzek, K. E., & Lee, K. M. (2009). The effects of competition on intrinsic motivation in exergames and the conditional indirect effects of presence. *Proceedings of the 12th Annual International Workshop on Presence, Los Angeles, CA*, 1–8.

Song, H., Kim, J., Tenzek, K. E., & Lee, K. M. (2013). The effects of competition and competitiveness upon intrinsic motivation in exergames. *Computers in Human Behavior, 29*, 1702–1708.

Sony Computer Entertainment America LLC. (2010). Sports Champions. *San Diego Studio and Zindagi Games* [PlayStation 3]. Tokyo, Japan: Sony Computer Entertainment, Inc.

Staiano, A. E., Abraham, A. A., & Calvert, A. L. (2012). Motivating effects of cooperative exergame play for overweight and obese adolescents. *Journal of Diabetes Science and Technology, 6*(4), 812–819.

Velez, J. A., & Ewoldsen, D. R. (2013). Helping behaviors during video game play. *Journal of Media Psychology: Theories, Methods, and Applications, 25*(4), 190–200.

Vorderer, P., Hartmann, T., & Klimmt, C. (2003). Explaining the enjoyment of playing video games: The role of competition. *2nd International Conference on Entertainment Computing*, 1–9.

Waddell, J. C., & Peng, W. (2014). Does it matter with whom you slay? The effects of competition, cooperation, and relationship type among video game players. *Computers in Human Behavior, 38*, 331–338.

Author Biographies

Gerry Chan is a Ph.D. student in the School of Information Technology at Carleton University located in Ottawa, Canada. He has a background in Human–Computer Interaction (HCI) and psychology. His research interests are in information visualization, computer-aided exercise, and player modeling in digital games. Recently, he has been working on research involving the use of wearable technologies and gamification principles for encouraging a more active lifestyle.

Gerry is particularly interested in the social and motivational aspects of the gaming experience. He enjoys playing games because he believes that games are valuable learning tools and offer ways for building stronger social relationships with others. For example, a collaborative game such as *LEGO Star Wars* or a video game such as the *New Super Mario Bros* allows players to learn about each other's personalities and builds stronger relationships through gameplay.

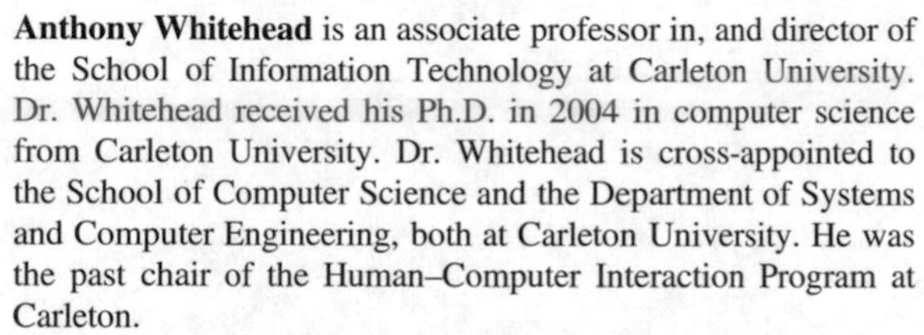

Anthony Whitehead is an associate professor in, and director of the School of Information Technology at Carleton University. Dr. Whitehead received his Ph.D. in 2004 in computer science from Carleton University. Dr. Whitehead is cross-appointed to the School of Computer Science and the Department of Systems and Computer Engineering, both at Carleton University. He was the past chair of the Human–Computer Interaction Program at Carleton.

His research interests can be described as practical applications of machine learning, pattern matching, and computer vision as they relate to interactive applications (including games), computer graphics, vision systems, and information management. In particular, he has been focussed recently on using sensors to facilitate a change in passive screen time to more active screen time and using procedural content generation (PCG) as a means to generate variety and encourage long-term engagement in active games and exergames.

Avi Parush is an associate professor at Technion, Israel Institute of Technology. With an academic background in cognitive experimental psychology, his areas of expertise are human factors engineering, human–computer interaction, and usability engineering. His professional and academic career in human factors has spanned over 30 years. He has devoted his career to influencing the design of workplaces and the tools people use to make their lives easier, safer, and more beneficial. He is internationally known as an expert in conducting usability studies. He is the founding editor in chief of the *Journal of Usability Studies* and the author of the Morgan Kaufmann book *Conceptual Design for Interactive Systems*. Teamwork in complex socio-technical systems, such as health care and first responders, is currently his main research focus, building on a long-standing passion for understanding and improving the relationship between people and technology.

Parush has explored gamification elements in simulation-based training and how they influence engagement and learning. He is currently studying the use of games on mobile devices to engage and empower young celiac patients.

FaceMaker—A Procedural Face Generator to Foster Character Design Research

Valentin Schwind, Katrin Wolf and Niels Henze

Abstract Understanding the effect of facial features on human's perception and emotion is widely studied in different disciplines. In video games, this is especially important to improve the design of virtual characters and to understand their creation process. Virtual characters are widely used in games, virtual therapies, movie productions, and as avatars in e-commerce or in e-education. Studying the design of virtual characters is challenging as it requires to have tools at hand that enables the creation of virtual characters. Therefore, we developed a system that enables researchers to study the design process of virtual faces as well as the perception of such faces. We developed a 3D model of the Caucasian average face and implemented design parameters that can be manipulated to change the face appearance. We integrate the face creation system into a web application, which allows us to conduct studies in the large. The application has been validated through a cluster analysis of procedurally generated faces from 569 participants which created 1730 faces.

1 Introduction

Virtual characters are commonly used in games, virtual therapies, movie productions, and as avatars in e-commerce or in e-education solutions. Due to the recent success of virtual reality and head-mounted displays, the importance of virtual characters will likely further increase over the following years. The success and the acceptance of an avatar in a movie, a game, or a therapy highly depend on how the

V. Schwind (✉) · N. Henze
University of Stuttgart, Stuttgart, Germany
e-mail: valentin.schwind@vis.uni-stuttgart.de

N. Henze
e-mail: niels.henze@vis.uni-stuttgart.de

K. Wolf
HAW Hamburg, Hamburg, Germany
e-mail: katrin.wolf@acm.org

© Springer International Publishing AG 2017
O. Korn and N. Lee (eds.), *Game Dynamics*,
DOI 10.1007/978-3-319-53088-8_6

character is perceived. In particular, the general appearance, the facial features, and the character's emotions influence the reaction of the player, customer, audience, user, or patient. Despite a body of work from different disciplines, the effect of a character's appearance is not completely understood.

Furthermore, self-customization and individualization of personal avatars are commonly used mechanisms to strengthen the connection between users and their virtually created characters. Game developers know that this connection intensifies the immersion in virtual worlds. Today, avatar creation systems are major aspects, especially at the beginning, of current role-play games (RPGs). A lot of recent game series or titles (Fallout, The Elder Scrolls, World of Warcraft, Mass Effect, The Sims) make use of individual customizations of virtual avatars to increase the emotional relationship between players and the game.

Avatar generators are also used in a non-gaming context. Virtual chat applications such as in AltspaceVR, Twinity, or at IMVU.com offer customized appearances and styles of virtual and help users to express themselves and to communicate their personal attitudes. With virtual clothing e-shopping assistants and virtual fitting rooms, such as clothing techniques presented by FITLE.com, users can scan their own bodies using their smartphone and customize their virtual self with garments before buying them.

However, little is known about how users create a virtual appearance of procedurally generated faces. To determine preferred characteristics of virtual faces, we conducted a user study in the large, based on an avatar generation system, which we called faceMaker. These results were already presented in a foregoing study (Schwind et al. 2015). In this article, we explain the system in detail and present the results of a further validation study of analysis using cluster algorithms. This demonstrates how the system can be used to extract user data and draw conclusions. We show how the system works and how it can be extended through further research. Thus, our contribution delivers a new kind of open-source tool at hand for game developers and researchers to understand how people create and perceive their own virtual creations.

Our work aims to provide a research tool that enables researchers of various VR domains, such as games, education, and therapy, to investigate effects of facial features and the look and appearance of a face on the perception of the virtual character, such as their influence on UX and emotions or even on the effect on the acceptance of an application. Similar to state-of-the-art role-play games (RPGs), our tool enables users to personalize virtual faces through providing a rich variety of parameters and options. We created faceMaker, a browser-based and real-time face generation application to provide the research community with a tool to investigate virtual faces in laboratory studies as well as in the large. We enable individual customizations of virtual faces to conduct reliable studies using a unified and complete unbiased avatar face model. Moreover, our web-based tool can be used as research apparatus in participatory studies where participants use our tool and researchers analyze the face design and measured results.

In this paper, we present main features of that system and explain how it can help to discover new findings in the perception of avatars and consequently in human–avatar interaction. The main contributions of our systems are (1) an open-source avatar creation system based on the average face, (2) a web-based user engaging a research-in-the-large tool, (3) the complete recording system of user activities and facial preferences, and (4) a back-end for data exploration and aggregating results.

2 Related Work

The connection between the appearance of a virtual character and ways of its customization through users is a research topic affecting many disciplines while being underexplored. For example, it has been found that physical attractiveness positively influences the opinion about other properties of a person (Dion et al. 1972; Eagly et al. 1991). Moreover, it is widely accepted that childish facial features, such as big eyes, trigger sympathy and let us find a character attractive (Langlois et al. 1991). In contrast, artificial characters, as robots, especially those who try to achieve human likeness are perceived as uncanny (Mori et al. 2012). These aspects can be transferred to virtual characters and often depend on factors as aesthetics and cultural context (Schwind 2015). However, it is currently unknown how customization and individualization of characters influence these aspects.

In human–computer interaction research and psychology, recent work is especially interested in the use of character creation systems to learn how virtual characters are perceived and how they are designed. Chung et al. (2007), for example, suggest that the process of avatar creation leads to a stronger sensation of their cyber-self-presence and psychological closeness to their customized character. This is supported by a study of Bessière et al. (2007), which indicates that a player's self-customized character in World of Warcraft leads to more favorable attributes than their own self-rated attributes, especially for people with a lower psychological well-being.

Avatar generation systems can also help to examine and identify differences between certain groups of players: Heeter et al. (2008), for example, observed teens in designing games over three years and found that females use a very high level of avatar customizations, while males rather use predefined characters. A study by Rice et al. (2013) indicates similar interests in customization of an avatar between different age groups. Ducheneaut et al. (2009) compared avatar creation systems for three different virtual environments. Their study shows that users emphasize the capability to change body shapes, hairstyles, and hair colors.

Studying virtual characters and especially their design requires character creation tools. Developing usable character creation tools can be challenging and requires significant effort. Various demands on face model and on the system, like visual quality, adaptability, extensibility, and a neutral base model, impede the fast development of reliable avatar creation systems.

Apostolakis and Daras (2013) developed the reverie avatar authoring tool RAAT with the aim to provide a research tool that allows individual customizations of whole-body characters. The online tool is based on a JavaScript library that helps developers and researchers to address avatar feature requirements. One of the application's features is a server-side module, which allows users to automatically incorporate mesh and texture of their own face on an avatar. RAAT supports interactive whole-body customizations like clothes, hairstyles, and changing face textures but no parametrized morph blendings of a neutral face model like the average face.

3 Parametric Average Face

3.1 Model Requirements

One aim of our system is to offer a solution that enables researchers to understand how people design characters and how the designed characters look like. For reliable investigations, the initial state of an avatar face should meet the following requirements: (1) The face should contain a minimum of characteristics which could be preferred or rejected by participants. (2) The model should have a balanced and uniform distance from typical facial proportions for procedural changes. (3) The start face must originate from the surveyed population (in our case Caucasoids). (4) The start face should not include any additional or biasing content.

To fulfill the four requirements, we developed a model of the Caucasian average face which is widely used in anthropomorphism and attractiveness research, e.g., (Langlois et al. 1994; Rikowski and Grammer 1999). Previous work primarily relied on image compositions of photographs. However, composing multiple images of human faces removes skin details and leads to unrealistic facial symmetry. The psychological research found that especially these two properties lead to higher attractiveness (Grammer and Thornhill 1994). To enable researchers to determine when people try to add (or to remove) realistic features, we decided to restore these properties and to introduce the skin details parameter—a realistic skin texture with asymmetrical and irregular skin details. As the average human face has any individual characteristics, researchers are now able to implement additional parameters for their experiment and minimize unwanted effects.

3.2 3D Face Model

The base face of our system is a 3D model of the Caucasian average face aiming for a neutral appearance as starting point for individual parameter configuration of facial features. Based on these parameters, corresponding 3D models can be derived. Thus, the generating system enables morphings of the model as well as blendings of texture maps to customize the appearance of the face.

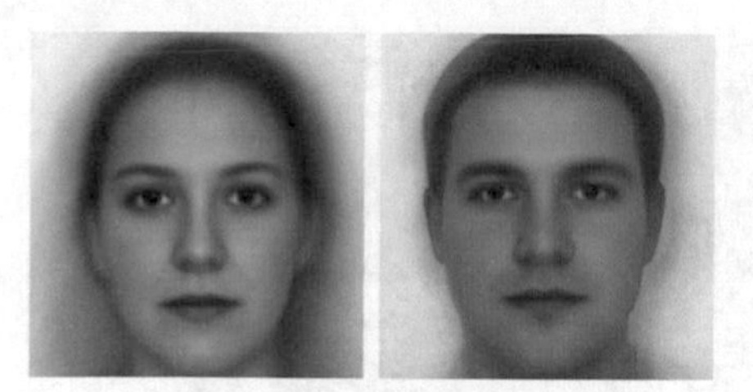 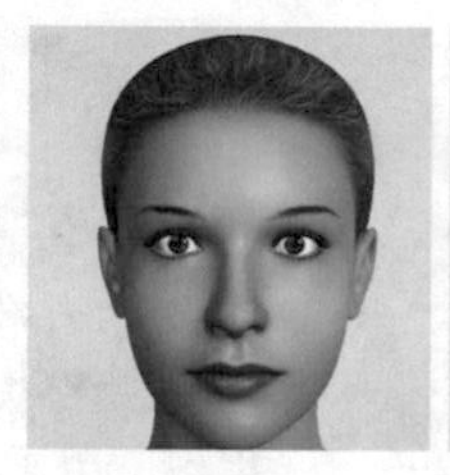 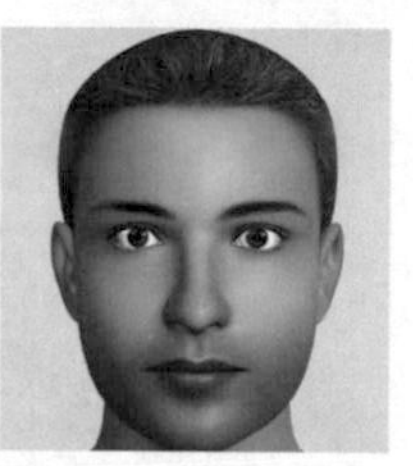

Fig. 1 *Both images left* Image compositions of the female and male average face, *both images right* average 3D face models

To address gender-related differences between females and males, we developed develop two separate average faces. The result based on image compositions including a large number of frontal images of neutral human faces. Images of 117 males and 151 Caucasian people from 18 to 40 years were retrieved from the online Face Database of the Parking Aging Mind Laboratory[1] (PAL) Database and from 3d.sk.[2] Children and older adults were not taken into account for the construction of the face model. This should be considered in further revisions or branches of the system. We used the automatic morphing method of PsychoMorph[3] developed by Tiddeman et al. (2001) to compose both average faces of adults (see both images on the left of Fig. 1).

The average faces were used as input for the PhotoFit feature of FaceGen[4] to create a first 3D model of the face. However, irregular polygon sizes, small artifacts, triangles that avoid subdivision smoothing, as well as the low image resolution of the calculated texture, were not useful for our purposes. Therefore, both meshes were retopologized, subdivided, and retextured using Autodesk Maya™ 2014 and Mudbox™ 2014 by two experienced CGI artists. The final results are shown in both renderings at the right in Fig. 1.

Physical attractiveness based on facial symmetry and golden ratio is considered as a result of averaging faces (Grammer and Thornhill 1994; Langlois et al. 1994). In order to determine whether the generated average faces met the assumption of facial symmetry and golden ratio, we applied Stephen Marquardt's φ-Mask as suggested by Prokopakis et al. (2013). The mask was developed to determine physical attractiveness and to determine deviations of facial symmetry. Thus, we assume that the φ-Mask can be applied to the generated 3D average faces. Figure 2 shows how Marquardt's φ-Mask is applicable to the female as well as to the male 3D average face.

[1]Parking Aging Mind Laboratory Database: http://agingmind.utdallas.edu/facedb.

[2]3d.sk Database: https://www.3d.sk.

[3]PsychoMorph Software: http://users.aber.ac.uk/bpt/jpsychomorph.

[4]FaceGen Software: http://www.facegen.com.

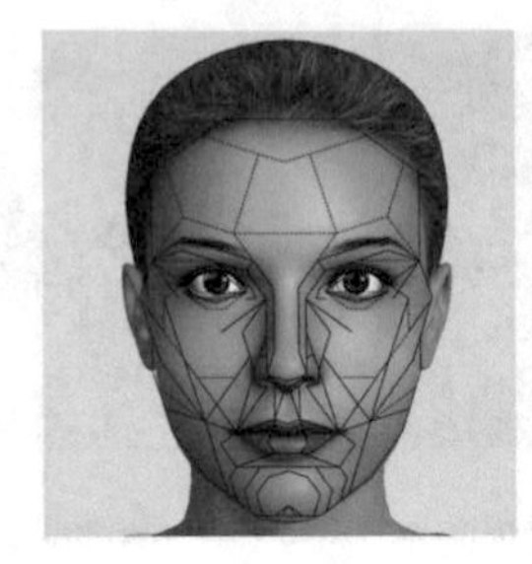
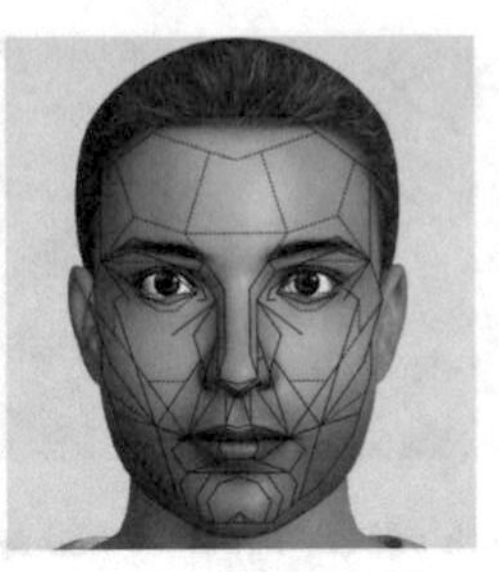

Fig. 2 Facial graph of Marquardt's golden ratio and symmetric φ-mask applied on both average faces

3.3 Parametrized Morphings

The female, as well as the male average face, can be continuously morphed using the face gender parameter. Additionally, according to our research questions investigated in our previous work (Schwind et al. 2015), we introduce three common character parameters: skin brightness, face style, and hair color. Face details were introduced to counteract the resulted skin smoothness of the average face.

To get an impression of prevalently used facial customization techniques, we examined the avatar customization systems of 9 commercial RPG: Mass Effect III, The Elder Scrolls: Oblivion/Skyrim, The Sims II/III, World of Warcraft, Destiny, Dragon Age I/II. Similar parameters (e.g., nose length and size) were aggregated, and inappropriate features (e.g., tattoos, scars, elf ears) were not taken into account. Due to their complexity and lack of parameterization, additional hair styles were not developed yet. Because of the limited space of a user interface and because all the parameters should be treated and distributed equally, we chose a maximum of 32 parameters. Table 1 shows all implemented parameters, their impact from left (−100%) to right (+100%) and the default value (*). Every parameter corresponds to a certain target morphing model (m) or texture blending (t). All morphing targets were modeled by hand. Model changes were only performed by vertex transformations in areas of the corresponding facial region (Fig. 3).

Morphings (also known as blend shapes or vertex displacements), as well as texture blendings, are used to change the appearance of the average face and to allow multiple parameterized customizations at the same time. In order to structure all parameters meaningfully, we developed a classification scheme that groups adjustments in eight facial domains (c.f. Fig. 3): common face parameters, eyes, eyebrows, nose, outer face, jaw and cheeks, mouth and lips, and makeup. Interface groups and infographics of the final application were separated according to this scheme. Currently, the system does not support facial domains or blendings. However, asymmetric surface details can be added using the skin detail texture blending, which includes no symmetries between the left and right half of the average face.

Table 1 All facial parameter scales

Parameter	–	Values	+	Type
Face gender	Female	Androgynous*	Male	tm
Face style	Realistic*		Cartoon	m
Face details	None	Half*	Full	t
Skin color	Black	Average*	White	t
Hair color	Black brunette	Med. Blonde*	Red bright blonde	t
Eyes color	Black brown*	Amber blue	Lt. Blue Green	t
Eyes shape	Droopy down	Round oval*	Almond up asian	m
Eyes opening	Narrow	Average*	Wide	m
Eyes size	Small	Average*	Big	m
Eyes height	Up	Average*	Down	m
Eyes distance	Narrow	Average*	Wide	m
Eyes orbit	Bulgy	Average*	Cavernous	m
Eyes rotation	In	Average*	Out	m
Eyebrows color	Black brunette	Med. Blonde*	Red bright blonde	t
Eyebrows shape	Pointed straight	Average*	Round hooked	m
Eyebrows strength	Thin	Average*	Thick	t
Nose shape	Snub	Average*	Hooked	m
Nose length	Short	Average*	Long	m
Nose width	Thin	Average*	Thick	m
Nose bridge	Thin	Average*	Thick	m
Nose cartilage	Round	Average*	Flat	m
Forehead size	Down	Average*	Up	m
Ear size	Small	Average*	Big	m
Throat size	Thin	Average*	Thick	m
Jaw shape	Triangle	Average*	Squared	m
Jaw length	Long	Average*	Short	m
Chin shape	Pointed	Average*	Cleft	m
Cheeks shape	Full	Average*	Scraggy	m
Lips volume	Thin	Average*	Full	m
Lips size ratio	Upper lip	Average*	Lower lip	m
Mouth shape	Down	Average*	Up	m
Mouth width	Wide	Average*	Narrow	m
Mouth height	Up	Average*	Down	m
Mouth depth	Backwards	Average*	Forwards	m
Makeup eyes shadow none*	None*		Full	t
Makeup lipstick	None*		Full	t
Makeup rouge	None*		Full	t

*default value, *t* texture blending, *m* mesh morphing

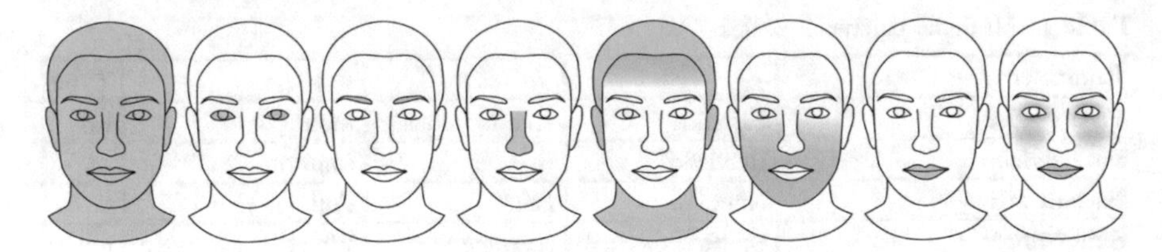

Fig. 3 Area classification used for vertex selection and infographics of the final application

4 FaceMaker Implementation

A system that allows manipulating the human average face model should meet the following criteria: (1) Reach a large number of participants under similar conditions as when players usually build avatars (at home). (2) A contemporary and interactive rendering engine, which is able to blend all morphings. (3) An easy to understand user interface, which could be randomized for every session to avoid sequence effects. (4) Provide support that helps to complete certain objectives and questionnaires.

In order to meet all requirements, we developed faceMaker. The browser application was developed to control the facial changes of the average face using parametric values. Our application design prevents the calculation of mean characteristics. First, multiple features on a single scale allow no reliable assumptions about a new means. And second, participants are not forced to change any values. However, to understand the concepts of different faces and to ensure that participants are able to create them, the 6 following objectives were introduced as design tasks: (A) A personal arbitrarily avatar face; (B) an uncanny, repulsive face; the stereotypical, positive-related face of an attractive (C) heroine; and (D) hero; the stereotypical face of a (E) female villain; and (F) male one. Therefore, we assume that faces of the category A, C, and D rather evoke positive feelings and that faces of B, E, and F are rather related to negative associations.

4.1 Online Face Generator

The application was developed with HTML, JavaScript, and jQuery. MySQL and PHP were used for client–server communication. For the implementation, WebGL is used as a rendering engine.[5] Hardware acceleration of more than 3 morphing targets simultaneously is currently not supported by Three.js. Therefore, we implemented a CPU-based software algorithm for multiple blendings of morph targets. Three directional lights and a slight ambient light in neutral white are used

[5]ThreeJS, WebGL Engine: http://threejs.org.

for lighting. The key light casts shadow maps. The orbiting camera can rotate within 180° in front of the face by clicking the left mouse button. The face consists of the head model, eyes, and eyelashes. No animations were added to the face. A gray t-shirt was added for a neutral transition from décolleté. The application runs in full screen of a browser window. The background is dark gray. Anti-aliasing is enabled.

For statistical and usability reasons, we decide that all parameters should be controlled with linear scales starting from a neutral point which corresponds to the human average face. This was realized using sliders but led to the disadvantages that participants sometimes will not change the default value. This can cause bias to the center, which is not intended in some studies or results in the need for a larger sample to find significant differences. Nevertheless, we decide that participants should not be forced to do something they do not want to change or should fix randomized parameter values they do not know.

All facial parameters can be controlled using horizontal sliders. The pixel width of each slider is 200. To the left and to the right, each tenth segment is highlighted with a stepped marker. No cursor snaps are used. Each parameter and all group boxes are labeled with the name of the facial parameter or domain. Additional states (e.g., colors) between left and right are labeled using tool tips below the slider. Each parameter can be set back to the default value with a button at the right. The majority of parameter sliders is gray. Only color changing sliders are equipped with color scales according to their parameter range. For every new user, all sliders within a domain group as well as the group box itself are randomly distributed on the left or right column of the browser. A help icon for each group box opens a help description, which informs the user about the corresponding changes.

4.2 Requirements and Compatibility

The system currently supports the browsers Firefox, Chrome, and Opera and offers multi-language support. An internet connection is only required while loading the application and submitting a face. The system needs a PC or mobile device with hardware-based 3D graphic acceleration to run smoothly. We recommend a graphics card with the performance of a NVIDIA GeForce GT 650M or higher.

4.3 Measurements, Back-End, and Extensions

The application is able to detect a couple of measurements: processing times, facial changes, resets, self-assessments, and facial parameters. All measurements are saved in cookies and are stored in the server's database after submitting a face. To filter for demographics, tasks, to view faces, and to save aggregated samples of certain face groups, we developed an application back-end. Data from here can be

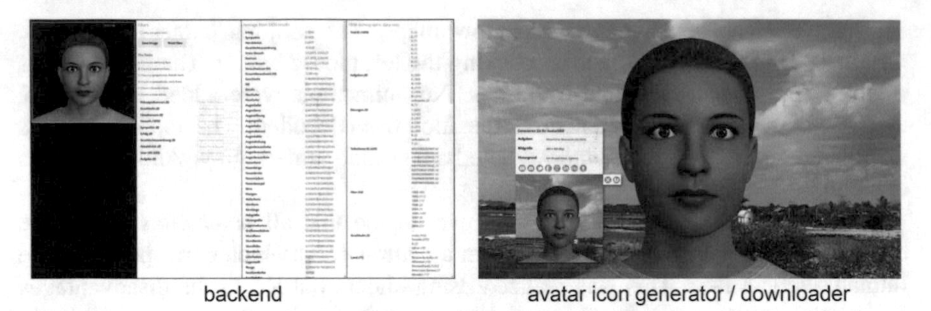

Fig. 4 Screenshots of the faceMaker back-end for evaluation purposes and icon generator/model downloader tool

used to view the faces participants create and to see which parameters they prefer. For example, the back-end can be used to see faces created by different genders, age groups, or country. The tool is also able to aggregate faces in a certain degree of perceived realism, attractiveness, or likeability. Graphical bars emphasize averaged or summed-up values behind each number.

Our system also includes an avatar image creation tool for saving the generated images to motivate people to use faceMaker. The small editor opens access to all faces in the database. Users can now change the background of the image and render each face from any perspective in a certain image size (optimized for Skype, Twitter, Gravatar accounts). The image can be shared via Facebook, Twitter, Google Plus, and LinkedIn. The tool also enables users to save the faces as OBJ and to download skin, hair, and eyes textures. Our tool can easily be combined with standard web surveys, which allows gathering additional information. Such combination of the user-designed faces and questionnaire inputs provided by faceMaker opens up a good platform for investigating virtual characters and for better understanding how their appearance is influencing the perception of an avatar at the first place, and the resulting effects on the acceptance of applications in a second place (Fig. 4).

4.4 Delimitation from Avatar Generators in Games

Our avatar generation system differs in many respects from avatar generation systems in games. For example, our system only supports human-like characters. Body decorations (e.g., tattoos, jewelry), aliens, or fantastic creatures (elves, orcs, etc.) are not supported. Thereby, our system differs from avatar generation system in commercial game series like Mass Effect, The Elder Scrolls, or Fallout. Another difference is animacy during the creation process. Avatar creation systems in games as in The Sims present their character in an interactive style with animated postures while changing parameters. Due to limited resources in a browser-based system, we only apply render updates to the render engine when a user triggers changes. Due to the lack of animation of the avatar face, which has to be rendered all the time, our system does not continuously consume system resources. Thus, faceMaker runs on devices in battery mode over a long period of time.

Another difference between avatar generation systems and faceMaker is the structure of the graphical user interface. To avoid biases, the parameter sliders of faceMaker are randomly arranged, which is not the case in games. Here, we found the most important settings (gender, race, etc.) at more accessible places in the GUI than, for example, finer adjustments of the mouth. Another distinguishing feature between GUIs in faceMaker and game is that faceMaker only use faces not the rest of the body, hair styles, or clothes.

5 Cluster Analysis and Statistics

To validate faceMaker, we conducted two studies: In a foregoing study, we validated the practical use of faceMaker in "the wild" through determining the preferred characteristics of stereotypical avatars. The results (14) show that it is possible to recruit a larger number of participants with faceMaker and to draw conclusions about the user's concepts and preferences of virtual avatars. We compared arbitrary faces with other categories. The results of this study finally showed that people rather prefer to create attractive female heroic faces than other faces.

As a second analysis, which we present below, we conducted a clustering analysis to understand which kind of faces was created without considering the objectives directly. We assume that our objectives deliver a higher contrast between different kinds of faces and relieve clustering without using any tasks for participants.

5.1 Procedure

The application procedure includes five steps: (1) A session starts with demographics including gender, age, origin, consummation of games and movies. Furthermore, we asked them to accept the terms of use. (2) The application starts and a pop-up window appears where a participant receives the instructions. After confirming the instructions, a participant could use the 37 sliders to change the average face. (3) Before submitting, a participant had to fulfill four assessments, which were not considered in the cluster analysis. (4) The application reloads after submitting and goes back to step 2. (5) After submitting 6 faces, a participant could view and download the generated faces. Cookies ensured that each task could only be performed once and repeated after the sixth trial. Cookies were deleted after 7 days. To avoid sequence effects, we used a balanced Latin square design starting with the face type which has the fewest participants.

5.2 Participants

We collected the results of 569 participants (313 males, 247 females, 9 n.a.) who created 1730 faces. The mean age was 30.58 (SD = 11.86). The participants were mainly recruited via mailing lists, social networks, and advertisements. 127 (22%) participants pointed out that the play games daily, 108 (19%) more than once a week, 68 (12%) once a week, 40 (7%) once a month, 125 (22%) play infrequently, 101 (18%) never. A total of 192 (14%) participants pointed out that they watch movies every day, 192 (34%) more than once a week, 171 (30%) once a week, 53 (9%) once a month, 62 (11%) watch infrequently movies, and 10 (2%) never. The median browser resolution on a participant's display was 1708 × 925. A total of 291 (51%) participants used Firefox, and 278 (49%) used Chrome as a browser.

5.3 Clustering and Multi-dimensional Scaling

To group and visualize the procedural faces systemically, cluster analysis was employed. Among different clustering algorithms, we applied the expectation–maximization (EM) algorithm for several reasons: EM does not require a predefined number of cluster (as kmeans), and the method iteratively searches for the maximum likelihood and was adopted in the previous work, e.g., for face detection as used by Rujirakul et al. (2014). It is known, that different cluster algorithms often produce inconsistent results. Different distance measurements can be used to validate the consistency of an algorithm. We decided to use the Euclidean distance metric because other metrics such as Ward's method tend to establish equal cluster sizes. To understand how the clusters are related to each other, we developed a spatial map to visualize the results. The distance matrices were used to conduct a multi-dimensional scaling (MDS). This approach cannot provide the accuracy of a face classification or the complete clustering; however, vividly illustrates the similarities or differences of faces. EM clustering was conducted in Weka, distance metrics, and multi-dimensional scaling was computed using R cmdscale.[6]

5.4 Cluster Analysis and Statistical Results

Based on the training set only given by the created facial parameters, EM clustering delivers 6 nodes: Cluster 0: 282 (17%), Cluster 1: 362 (21%), Cluster 2: 169 (10%), Cluster 3: 209 (12%), Cluster 4: 229 (13%), Cluster 5: 457 (27%). The 6 objectives that people created were verified using a cluster assignment. Cluster 0 was assigned

[6]cmdscale—A R-library for multi-dimensional scaling: Retrieved June 2016 from https://stat.ethz.ch/R-manual/R-devel/library/stats/html/cmdscale.html.

to F (male villain face), Cluster 1 to C (attractive heroine face), Cluster 2 to A (arbitrary face), Cluster 3 to B (repulsive face), Cluster 4 to E (female villain face), and Cluster 5 to D (male hero–villain). We would like to note that the results of the EM algorithm and the assignment procedure are incidentally identical to the number of objectives. The parameter results of the cluster analysis were added to the plotted diagram of the multi-dimensional scaling (see Fig. 5). 55.0% of the faces were incorrectly clustered instances. The class attribution table (not illustrated) reveals, that Cluster 1, for example, shares 257 instances with objective A and C, which could be explained by the tendency of participants to create faces in the arbitrary task that are similar like in the female hero task.

The plotted MDS map (Fig. 5) reveals a dense filament including two main consolidations at the top and clusters of faces that people prefer to create. Parametric values of the cluster centers were used to render faces and were placed on the MDS map. Two main clusters are connected in the main filament structure. Cluster 1, 2, and 4 as well as the average faces of objective C and E are on the female "side" on the map. Cluster 0 and 5 as well as faces of objective D and F are on the male "side" of the map. Since all parametric changes were weighted equally, the arrangement and the results of this cluster analysis reveal that the sum of parametric changes is made according to gender and appeal. The arbitrary face (A), which participants created without any restrictions, is very close to the female cluster of faces. We also see that the cluster center of female villains (4) is much closer to female heroines (1) than the male villain center (0) from the male heroes one (5). The cluster center of the repulsive face is outside from the cluster centers of male or female faces.

The MDS map of the cluster analysis shows that facial properties of the created faces depend on gender and appeal. Male faces tend to be placed at positive x-values; female faces are rather placed at negative ones. Appealing faces (heroes) tend to have positive y-values, and not appealing (repulsive faces) have negative y-values. Using this approach, we are now able to derive stereotypical faces and avatars from procedural ones. For example, to find very female facial properties, we can look at samples very close to the cluster centers 1 and 2. In contrast, very repulsive faces are outside of the main cluster of females and males. The repulsive face samples show a very distributed pattern. This can be explained by the fact that repulsive faces have no certain features or patterns. However, they differently deviate from the human ideal and human average proportions, which people try do not violate while creating appealing faces.

Besides the results of the cluster analysis, we looked deeper into the statistics of the 37 parameter scales. Figure 6 shows how often people interact with these parameters. The bar charts show the average values of parameter changes, resets, and views (instead of using the orbit camera). Face gender, hair color, eye shape, and skin color are changed very often in contrast to ear size, rouge, lip stick, and mouth height. The resets indicate which parameters were often reset to the human average. The most resetted parameters are related to the eyes. That were in particular eye depth, eye distance, and eye rotation. The parameter views in Fig. 6 show which parameters were often visited by the mouse cursor. Hovering a slider

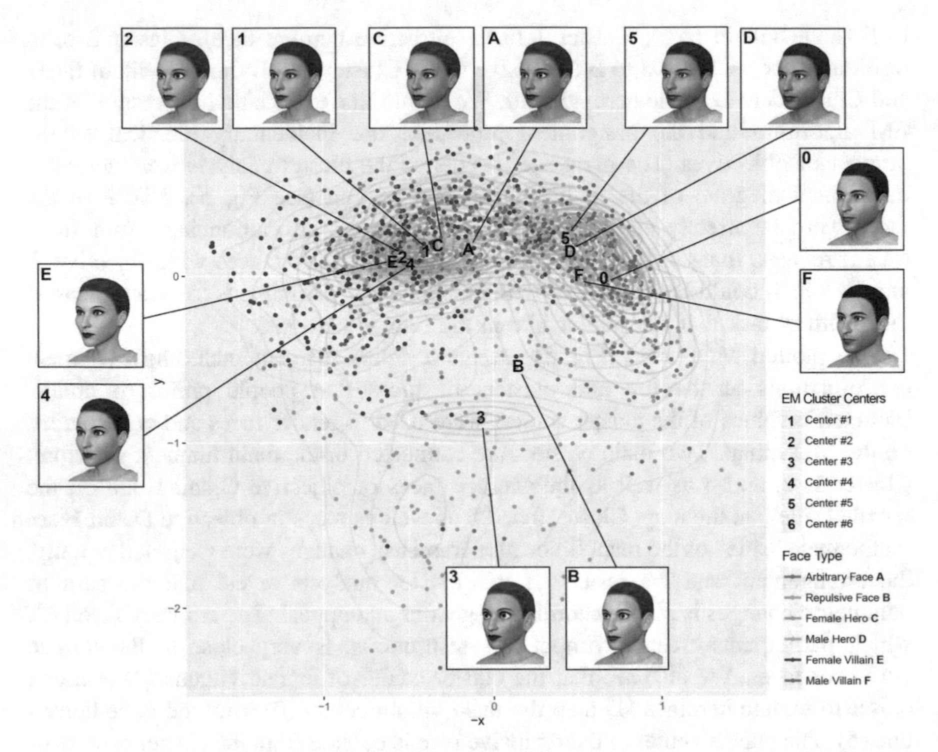

Fig. 5 Plot of the MDS analysis including 1730 faces from 569 participants and location of the average faces (*A–F*) resulted from the users' objectives and the 6 cluster centers (*1–6*) delivered by expectation–maximization (EM)

zooms or rotates the camera perspective to a certain region. The diagram shows on which facial regions users seem to be interested in while creating avatars.

6 Discussion

In this study, we present the results of a cluster analysis based on 1730 procedural faces created by 569 participants which used the online avatar generator faceMaker. To validate faceMaker, we used EM clustering. We found the same amount of cluster nodes as given objectives (arbitrary face, repulsive face, female face, male face, female villain, male villain). Through cluster alignment, we determined which cluster generally corresponds to the stereotypical average face. Plotting the results of a MDS analysis reveals how the faces are related and which face types rather deviate or correspond to the human average and human ideal. The map reveals two main distributions of male and female faces and shows how participants create faces according to their concept of stereotypes. Class attribution reveals that Cluster 1

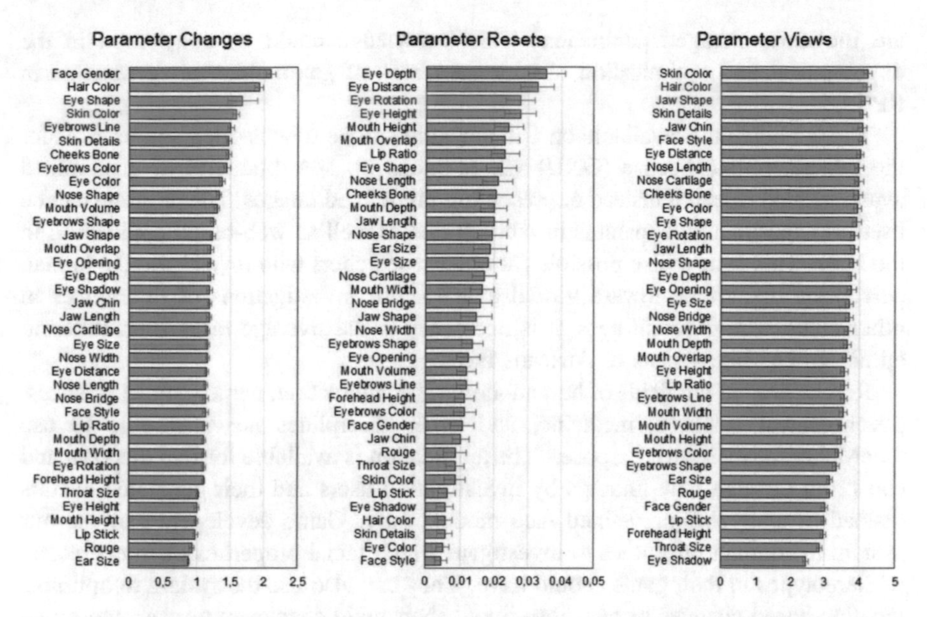

Fig. 6 Bar charts of parameter changes, resets, and views. *Error bars* show standard error (SE)

shares 257 instances with the arbitrary and female hero face. If having a choice, our participants are more willing to create female faces instead of male faces. Positively associated faces (female and male heroes) get smooth skin, realistic and attractive proportions, and natural average skin color. Villains get bright skin, distinctive cheek bones, and exaggerated jaws. Female villains receive strong lip stick, male villains get strong eyebrows. Repulsive faces get features that strongly deviate from the human norm, which is divided into two main distributions of male and female faces. They were exaggerated with unnatural violations against the human average. Thus, using these procedural types of faces, we are now able to look into preferred or not preferred stereotypical concepts of avatar faces and can deduce which facial properties are rather preferred or not preferred.

The avatar and face creation system faceMaker[7] enables researchers to conduct valid online studies reaching a large number of participants. In the cluster analysis, we revealed groups of faces and their relation to each other. We found how procedurally generated faces could be used to identify stereotypical faces and which parameters they possess. Game developers can now deduce which avatar faces are rather preferred or not preferred by users. Knowledge about preferred or not preferred facial characteristics of avatars gained by faceMaker can be considered in design decisions about the appearance of game characters and future avatar generators. The usage of parameters provided in faceMaker allows optimizing such avatar generators. For example, we showed that gender, hair color, and eye shape

[7]FaceMaker Online Application: http://facemaker.uvrg.org.

are the most changed parameters. This knowledge could be considered in the development and optimization of user interfaces of game character generators in RPGs.

The application is available on GitHub[8] and can be downloaded and used under the general public license (GNU) v2. It opens up new branches of customized avatar-related research instead of presenting predefined images. The program can be used as experimental apparatus in laboratories as well as web-based application in the large. This is now also possible for other researchers who investigate the human perception or attitude toward virtual avatars. For investigations of differences in other cultures or different ages, it is necessary to use average faces from different ethnic groups like Asians or Africans faces.

Researchers in the fields of human–computer interaction, games, social sciences, psychological sciences, medicine, and other disciplines are now able to use faceMaker for their own purposes. The application is available for free disposal and can profit through new findings by investigating users and their generated avatars instead of analyzing predefined face models only. Game developers can use our system to conduct prestudies to investigate which facial properties main characters or stereotypes in their game should have. They can also use the system to optimize location-based changes or questions about their main characters or user interfaces. Furthermore, we suggest future research about how users create procedural faces they already know (as celebrities), look like themselves, or investigate gender or cultural differences in the face creation process.

Acknowledgements Special thanks go to Katharina Leicht, Verena Dengler, Julia Sandrock, and Solveigh Jäger. We thank the German Research Foundation (DFG) for financial support within project C04 of SFB/Transregio 161. This work was also supported by the cooperative graduate program "Digital Media" of the University of Stuttgart, University of Tübingen, and the Stuttgart Media University.

References

Apostolakis, K. C., & Daras, P. (2013). RAAT - The reverie avatar authoring tool. In 2013 18th International Conference on Digital Signal Processing (DSP) (pp. 1–6). IEEE. 10.1109/ICDSP.2013.6622788

Bessière, K., Seay, A. F., & Kiesler, S. (2007). The Ideal Elf: Identity Exploration in World of Warcraft. CyberPsychology & Behavior, 10(4), 530–535. 10.1089/cpb.2007.9994

Chung, D., DeBuys, B., & Nam, C. (2007). Influence of Avatar Creation on Attitude, Empathy, Presence, and Para-Social Interaction. Human-Computer Interaction. Interaction Design and Usability, 4550, 711–720. 10.1007/978-3-540-73105-4

Dion, K., Berscheid, E., & Walster, E. (1972). What is beautiful is good. Journal of Personality and Social Psychology, 24(3), 285–290. 10.1037/h0033731

Ducheneaut, N., Yee, N., Wadley, G., Alto, P., & Alto, P. (2009). Body and Mind : A Study of Avatar Personalization in Three Virtual Worlds, 1151–1160.

[8]faceMaker at GitHub: https://github.com/valentin-schwind/FaceMaker.

Eagly, A. H., Ashmore, R. D., Makhijani, M. G., & Longo, L. C. (1991). What is beautiful is good, but … A meta-analytic review of research on the physical attractiveness stereotype. Psychological Bulletin, 110(1), 109–128. 10.1037/0033-2909.110.1.109

Grammer, K., & Thornhill, R. (1994). Human (Homo sapiens) facial attractiveness and sexual selection: the role of symmetry and averageness. Journal of Comparative Psychology, 108(3), 233–242. 10.1037/0735-7036.108.3.233

Heeter, C., Egidio, R., Mishra, P., Winn, B., & Winn, J. (2008). Alien Games: Do Girls Prefer Games Designed by Girls? Games and Culture, 4(1), 74–100. 10.1177/1555412008325481

Langlois, J. H., Ritter, J. M., Roggman, L. a., & Vaughn, L. S. (1991). Facial diversity and infant preferences for attractive faces. Developmental Psychology, 27(1), 79–84. 10.1037/0012-1649.27.1.79

Langlois, J. H., Roggman, L. A., & Musselman, L. (1994). What is average and what is not average about attractive faces? Psychological Science, 5(4), 8.

Mori, M., MacDorman, K., & Kageki, N. (2012). The Uncanny Valley [From the Field]. IEEE Robotics & Automation Magazine, 19(2), 98–100. 10.1109/MRA.2012.2192811

Prokopakis, E. P., Vlastos, I. M., Picavet, V., Nolst, G., Thomas, R., Cingi, C., & Hellings, P. W. (2013). The golden ratio in facial symmetry. Rhinology, 51(4), 18–21. 10.4193/Rhino12.111

Rice, M., Koh, R., Lui, Q., He, Q., Wan, M., Yeo, V., … Tan, W. P. (2013). Comparing avatar game representation preferences across three age groups. In CHI '13 Extended Abstracts on Human Factors in Computing Systems on - CHI EA '13 (p. 1161). inproceedings, New York, New York, USA: ACM Press. 10.1145/2468356.2468564

Rikowski, A., & Grammer, K. (1999). Human body odour, symmetry and attractiveness. Proceedings. Biological Sciences/ The Royal Society, 266(November 1998), 869–874. 10.1098/rspb.1999.0717

Rujirakul, K., So-In, C., & Arnonkijpanich, B. (2014). PEM-PCA: A parallel expectation-maximization PCA face recognition architecture. The Scientific World Journal, 2014. 10.1155/2014/468176

Schwind, V. (2015). Historical, Cultural, and Aesthetic Aspects of the Uncanny Valley. In Collective Agency and Cooperation in Natural and Artificial Systems (pp. 81–107). Cham: Springer International Publishing. 10.1007/978-3-319-15515-9_5

Schwind, V., Wolf, K., Henze, N., & Korn, O. (2015). Determining the Characteristics of Preferred Virtual Faces Using an Avatar Generator. In Proceedings of the 2015 Annual Symposium on Computer-Human Interaction in Play - CHI PLAY '15 (pp. 221–230). New York, New York, USA: ACM Press. 10.1145/2793107.2793116

Tiddeman, B., Burt, M., & Perrett, D. (2001). Prototyping and transforming facial textures for perception research. IEEE Computer Graphics and Applications, 21(October), 42–50. 10.1109/38.946630

Author Biographies

Valentin Schwind is Ph.D. student in the field of Human–Computer Interaction (HCI) under the supervision of Prof. Dr. Albrecht Schmidt at the University of Stuttgart. After completing his diploma in media computer science at the Stuttgart Media University (HdM), Schwind worked as freelancer and consultant for visual effects and game companies in Germany, Austria, England, and the USA. Later, he was research assistant and lecturer at the Stuttgart Media University and involved in research projects investigating virtual faces and avatars.

In 2015, he became researcher at the Institute for Visualization and Interactive Systems (VIS) of the University of Stuttgart focusing on the Uncanny Valley in Human–Computer Interaction and deploying metrics in the visual perception of virtual characters. He is now doctoral researcher of the collaborative research center in the SFB-TRR 161 investigating quantitative methods for visual computing. He published in renowned conferences and journals, received the best paper award at the conference Mensch & Computer 2015, and has been reviewer for leading conferences and journals.

Schwind teaches in the field of computer graphics and game development. His interests range from gaming, interaction, human perception, and cognition. In particular, he is interested in human–avatar interaction, virtual reality, and visual computing. He developed and supervised the creation of games, computer animations, and mobile applications including procedurally generated content.

Katrin Wolf is professor for Media Informatics at Hamburg University of Applied Science where she is teaching in the Media Systems program. Before she was professor for Media Informatics at BTK—University of Art and Design in Berlin. She was a postdoctoral researcher in the Human–Computer Interaction Group in the Institute for Visualization and Interactive Systems and the SimTech Cluster for Simulation Technology at the University of Stuttgart.

Wolf was a researcher and doctoral student at the Telekom Innovation Laboratories at the Technical University of Berlin, where she investigated ergonomics in mobile HCI. Her studies at the Berlin University of the Arts, where she received two German diplomas, provide her with a background in media, interface, and interaction design as well as with communication science.

Particularly, she is interested in human–computer interaction, interaction design, and using parametrically or procedurally generated content in games. Wolf published peer-reviewed journals and conferences. She organized scientific workshops, served as program committee member, and has been reviewer for leading conferences in her field. Wolf lectures interaction design as well as human–computer interaction in practical and theoretical courses for several years.

Niels Henze is assistant professor for socio-cognitive systems in the Institute for Visualization and Interactive Systems and the SimTech Cluster for Simulation Technology at the University of Stuttgart. Beforehand, he was postdoctoral researcher in the Human–Computer Interaction group at the University of Stuttgart. After receiving a diploma in computer science in 2006, he worked for European research projects at the OFFIS Institute for Information Technology. In 2008, he became a researcher and doctoral student in the Media Informatics and Multimedia Systems group at the University of Oldenburg. He worked for international projects and was responsible for teaching and tutoring. Henze finished his Ph.D. in 2012 with his thesis about Camera-based Mobile Interaction with Physical Objects under the supervision of Prof. Dr. Susanne Boll.

Henze's research interests are mobile–human–computer interaction and pervasive computing. Particularly, he is interested in large-scale studies using mobile application stores as a research tool, attention and smart notification management, as well as multimodal interfaces. Niels published in scientific journals and competitive conferences. He received awards from CHI, MobileHCI, and Mensch & Computer. He organized several scientific workshops, served as a guest editor for the International Journal on Mobile Human–Computer Interaction, and has been a reviewer for the leading conferences and journals in his field. Henze lectures human–computer interaction for several years. He developed and supervised the development of mobile applications to conduct large-scale studies that have been installed more than a million times. He is interested in using procedural generated content in games for conducting user studies in the large.

A Primer on Procedural Character Generation for Games and Real-Time Applications

Yanko Oliveira

Abstract Several of the hurdles in meaningful procedural generation are exacerbated when it comes to characters, especially when they are at the very center of the stage. Ontogenetic approaches (i.e., that try to mimic the desired result, instead of generating it by simulating natural processes) are usually the best fit for most applications, usually involving altering a base character and/or combining sets of modular parts. This is, however, still a design, art and engineering challenge. This chapter covers procedural character generation, while pointing at character design fundamentals that should be kept in mind when designing these systems. Starting by reviewing state-of-the-art techniques from games that have them as a main feature, it then covers mesh generation, altering existing meshes and scaling bones to generate diverse silhouettes. It also proposes a technique for 3D mesh combination that involves offline tagging of vertices in edges to create seamless joints between modules at runtime. Finally, it covers texture generation and combining texture channels in ways that allow even more varied characters.

1 Introduction

From a very early age, humans have the ability to recognize certain characteristics and behaviors of living beings, and cultural differences can even increase the chance of assigning living characteristics to inanimate objects (Inagaki and Hatano 2006). That is both a burden and an advantage when designing systems for generative characters. More so, our brain is trained "out of the box" to recognize faces (Haxby et al. 2002)—any simple combination of objects, arranged in a certain way, might spawn a character in our perception (Rieth et al. 2011). However, if something appears off in a face designed to be realistic, the illusion breaks as easily as it appeared. This can be generalized to whole characters in games: Having a

Y. Oliveira (✉)
deVoid Games, Hamburg, Germany
e-mail: yanko.oliveira@gmail.com

© Springer International Publishing AG 2017

O. Korn and N. Lee (eds.), *Game Dynamics*,
DOI 10.1007/978-3-319-53088-8_7

hyperrealistic model that animates poorly is much worse than a simple 2D drawing collage that has very simple animation, but holds the illusion well enough.

Although the uncanny valley was widely discussed and studied, there is still no definitive answer on how to avoid it completely (Schwind 2015): In the end, it is all about context and the suspension of disbelief; that is why thinking about function is extremely important when approaching procedural character generators.

In this chapter, some techniques for procedural character generation will be covered, expanding simple ideas from the 2D to the 3D world, revisiting some proven concepts in existing games and presenting new ideas on how to approach these challenges. Even though there is a lot of space to experiment with more abstract characters, the focus will be on modular approaches and how to extract as much as possible from these techniques. There will also be some examples using the Unity3D Engine as a base and code snippets in C#, but all can easily be ported to any other language and environment.

While a teleological approach (that attempts to simulate the natural processes that generate something) is very interesting for terrain, plants and other inanimate entities, characters in real-time applications such as games will usually follow the ontogenetic approach: observing the desired end results (that is, how characters should look like) and developing algorithms with these final results in mind. However, to design those algorithms, it is important to understand first what exactly makes a good character.

The most important thing for a character to be discernible from others and to have a strong visual impact is its silhouette—silhouette allows us to recognize which class a *Team Fortress 2* player is using in a fraction of a second (Mitchell et al. 2007), or easily tell *who's that Pokémon*. However, a good character also has to fit its context, the art style of the whole game, and its color palette can communicate a lot about it. These are all concepts that, if not taken into consideration, can ruin the best generational code.

2 State of the Art

In Relic's 2003 RTS *Impossible Creatures* (Relic Games 2003), the player can combine different animal body parts to create hybrid soldiers. Internally, the models are actually represented by groups of Bézier surfaces (usually referred to as *Bézier patches*), and all creatures need to be modeled with that technique (which is, to say the least, outside of a game artist's toolbox for quite a while now). Then, each body part would be tagged, and they needed to have exactly the same amount of termination points at the cross sections so that they could fit into one another. At runtime, the parts would be stitched together and a vertex-based 3d model created for rendering (Dunlop and Drew 2002). Even though not very practical for current engines and art workflows, this technique has some interesting implications: Because Bézier patches are meta-surfaces, dragging their control points automatically creates smooth topologies, and level of detail meshes can be generated on the fly.

Another solution for 3D models of arbitrary shapes is creating a mesh out of a point cloud or a similar way of generating implicit surfaces. This kind of technique is very common for terrain using voxels and variations of the marching cubes algorithm, but the same idea can be applied to characters. These methods are commonly referred to as "metaballs" when they are used for organic modeling.

Before the recent *No Man's Sky* (Hello Games 2016), *Spore* (Maxis 2008) was a game completely grounded on procedural generation with an even bigger self-imposed challenge: Creatures would be user-created. This required a system capable of results akin to those of a complete 3D modeling suite via an interface simple enough for any player to learn within a few minutes. Even eight years after its release, it is still one of the best resources on the topic.

To achieve this, the *Spore Creature Creator* combined modular body parts that would add up to a single generated implicit surface (by distributing several spheres whose radii were interpolated between user-defined control points) with what the developers referred to as "*Rigblocks*," which were modules that had different deformable targets created by baked-in animations (Choy et al. 2007). Chris Hecker's "*Liner notes for Spore*" (Hecker 2014) are a great source of information regarding these methods and highly recommended.

One of *Spore's* biggest achievements was that the characters would animate properly independently of their shape or limb quantity. For its animation pipeline, a tool was created where animators could work with a typical keyframe-based approach that would generate rig-agnostic animations, which would then be retargeted during runtime (Hecker et al. 2008). For games that are less ambitious, a particular art style might help: It is not uncommon to use funny-looking, simpler aesthetics and get away with less advanced animation systems. Games like *Gang Beasts* (Boneloaf 2015) and from the Dutch collective *Sokpop* (Sokpop 2016) are good examples.

There is a reason for games that put procedural characters front and center to be far and in-between: For proper, meaningful characters to emerge from algorithms, several factors must be kept in mind, and the technological and art hurdles are many. Building upon the knowledge gathered by the aforementioned games, we skip any possible one-dimensional beings and start with the next logical step.

3 Modular 2D Characters

A workflow for 2D characters can consist of simply layering parts together. Since all assets created will be directly represented on the screen (that is, there will be no rotation or change in the point of view), the first step is deciding on the purpose of the character and how it will be visualized: Is it a portrait? Is it full body? Does it have to animate? Can the animations be shared between all variations?

Fig. 1 Faces extracted from the *Papers, Please* trailer (Pope 2013b)

Concept art is an important tool at this step: Sketches of how the character would appear on screen can help determine which regions can be randomized from a set of pre-made parts. In Lucas Pope's *Papers, Please* (Pope 2013a), characters appear mainly from the torso up. All of them are generated by combining a base with head and shoulders with different sets of eyes, mouth and nose (Pope 2012). The animation for these portraits consists of fading in and out of view, and panning. Even with these very simple generational rules, the game can rely on constantly showing said characters on screen, without the feeling of repetition (Fig. 1).

In a more action-oriented game with full-body, animated enemies, several appendages, clothes, hairstyles or even whole body parts can be combined on top of a base character. That would allow complex animations to be shared between all generated characters, still keeping variation.

Modular procedurality in 2D is easier to approach, but can be limiting. Fortunately, combinatorics are on our side: Combining a handful of small sets can scale up to thousands of possibilities. The challenge, however, goes up a notch when it is taken up a dimension.

4 3D Characters

Several different techniques can be used for procedural 3D characters. The final purpose defines the approach: Experimental shapes and methods can be used for more abstract creations, but modular approaches akin to the ones for 2D characters mentioned above are more common. Again, form and method will be defined by function and intent. We begin with a simple example.

4.1 Procedural Chess Pieces: Creating Meshes from Scratch

X, a Game of YZ (Gitahy Oliveira 2015), is a chess-like game with procedural rules created for the 2015 *PROCJAM* (a yearly game jam that is highly recommended to all those interested in PCG) (Cook 2015). The game needed pieces that resembled regular chess pieces in some aspects, but could not have the exact same look to avoid confusion, as every one of them could move in completely different ways than their original counterparts.

The first step was dissecting what typical chess pieces have in common, visually. What immediately comes to mind is the difference between base, body and the "headpiece" that actually defines which piece is which. The classical design comes from lathe sculpting methods, where a piece of metal or wood is rotated by a motor and carved. In a 3D modeling software, the idea is replicated by rotating a spline around an axis. Because of the time constraints of a game jam, simpler approaches can be very valuable: Instead of having to implement a spline system and then a lathe operator that would generate meshes, creating a cylinder mesh and changing the radii of its segments would yield the same effect. Groups of segments were used for each one of the piece parts (base, body and headpiece), and each one of those had a different configurable set of parameters to make the final look possible.

In the game, each piece belonged to an archetype, so different configurations were created based on those: *Pawns* are plain and short, *Support* pieces are tall and have complex headpieces, *Heavy* pieces are wider, stubby and have a rook-like headpiece, and *Royal* pieces are pretty much the same as Support, but have a "panache," i.e., a small ornament over the headpiece. While developing the generator, more parameters were added with every iteration to suit the final look, including curves that modify the segments' radii, defining their overall shape. That takes away some interesting variations, but keeps pieces cohesive: The base of the pieces always tends to be wider at the bottom, while bodies vary a lot in the middle, but little where they join with the base and headpiece. Although this was not implemented in the game, one of several pre-made headpieces could have simply been attached to the end of the piece's body, making even more characteristic variations possible (Fig. 2).

4.2 Variations from Existing Meshes: Blend Shapes

Blend shapes are a very common solution for character customization in 3D games. The algorithm linearly interpolates the position of a given vertex in an origin mesh to the position of the equivalent vertex in another mesh—that is why for a proper

Fig. 2 Chess pieces generated by mimicking lathe sculpting

result, the number of vertices and their indexes needs to be the same between the base and target meshes. Because of that, the workflow for creating the models must keep in mind that the base model's edge loops need to be prepared for all the possible variations. Multiple target meshes can be used, and it is usually important that they do not "override" each other (i.e., if one target mesh moves the vertex -10 units and the other one +10 units in the X axis, the blended result will be as if there was no transformation to the base mesh).

Each interpolation between base and target mesh can be represented by a value in a [0,1] interval. Given a base mesh and a set of target meshes, the interpolations values for each target can be randomized to generate procedural characters. This can yield good results, and with target meshes that are different enough, a lot of variation is possible. The more extreme the target meshes, more noticeably varied results can be generated, but because all need to share the same vertex amount and have to come from the same base mesh, forcing extreme variations might not scale well on the art pipeline. However, for creating varied characters of the same kind (humans and orcs in an RPG, zombies on an FPS), this is usually the technique with the best cost–benefit relation (Fig. 3).

4.3 Proportions, Silhouette and Size: Rescaling Bones

As mentioned before, one of the most important factors for a character to be discernible from others and to have a strong visual impact is its silhouette. That means that if one wants to generate hundreds of different creatures for a player to discover, their shape will have to change drastically sometimes for them to keep being visually appealing. Changing proportions of individual body parts can

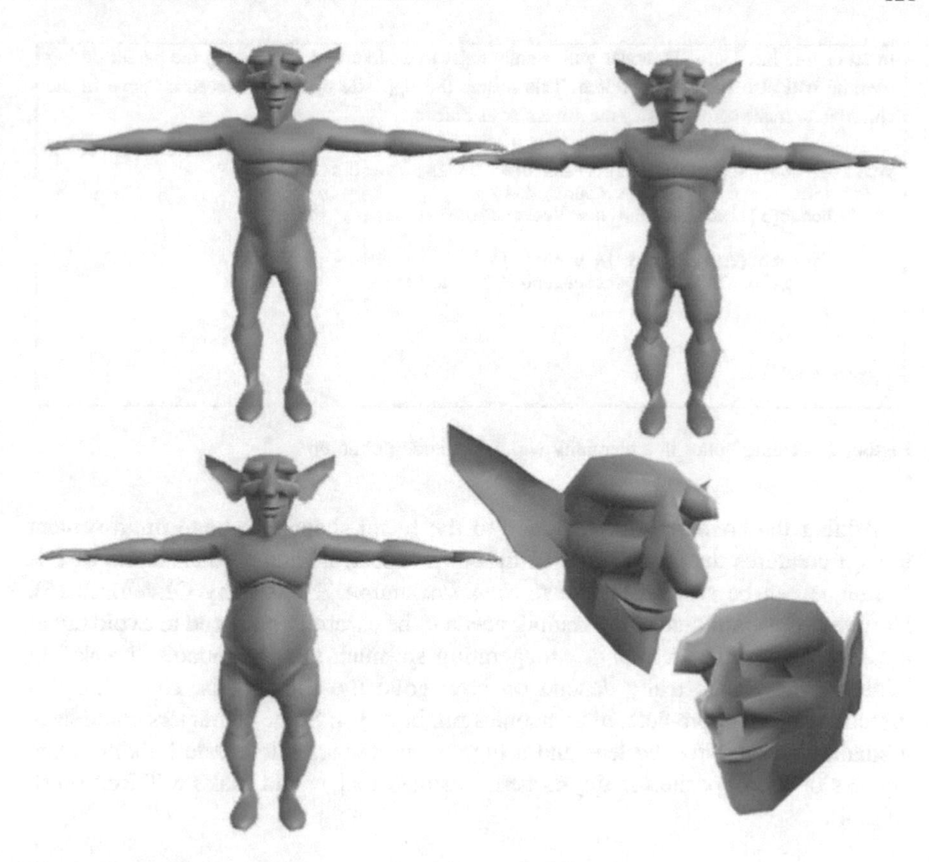

Fig. 3 Several blend shapes extracted from the same base mesh

completely transform a character's silhouette: compare, for example, the skeletons of a giraffe and a horse.

Character models for 3D games most likely will also have a skeleton rig to be animated, and we can take advantage of that to rescale the bones and deform the model in different ways. Obviously, leaving the scale completely random will generate undesirable results, so it is important to clamp the scale between minimum and maximum values that work well with the base model, a configuration that might even need to be adjusted on a limb-by-limb basis (Listing 1).

However, that leads to another problem: This algorithm makes legs shorter and longer, so some characters will either have their feet up in the air or under ground level. If it is possible to guarantee all the limbs end at the same height, one way to fix this is checking the distance from the bottom of one foot bone to the ground plane and offsetting the whole mesh by that distance—so this can be used easily for biped characters. For quadrupeds, however, a more robust IK solution might be necessary, since their front and hind legs might have different lengths and the whole pose of the body needs to compensate for that.

In an engine like Unity3D, bones will usually exist in a hierarchy, so resizing the parent during runtime will also scale the children. This means the opposite operation has to be done in the children, to make sure they stay the same size as before.

```
private float ScaleLimb(List<Transform> bones, float scale) {
  for (int i = 0; i < bones.Count; i++) {
      bones[i].localScale = new Vector3(scale, scale, scale);

      foreach (Transform t in bones[i]) {
          t.localScale = Vector3.one * 1 / scale;
      }
  }

  return scale;
}
```

Listing 1 Scaling bones in a hierarchy without affecting children

Adding the bone scaling technique to the blend shape is a huge improvement toward creatures that seem to be of different "species," and a small demo of this technique can be seen in the *Bestiarium Invocation Toy* (Gitahy Oliveira 2016). However, the amplitude of the scaling needs to be carefully balanced to avoid either creating very similar results or exaggerating so much that the models "break." In addition, the results really depend on how good the skinning is. Exploring this direction, different pre-defined archetypes might be set: Some characters could have a smaller variance for the legs and a bigger one for arms to create hulking, alien gorillas or the opposite for tiny-armed creatures that would make a T-Rex blush (Fig. 4).

4.4 Modular 3D Characters: Combining Meshes

It is relatively easy to follow the same structure as the 2D example of combining body parts in 3D models, being this mostly an art workflow challenge. However, simply putting objects together creates visual seams. That is why modular assets look great out of the box on clothed characters (it is easy to hide a seam for hands in shirt sleeves, for the torso in the pants, etc.). Adding appendages like horns, antennae, hats and hairstyles also easily creates variation by simply positioning different meshes on top of a base character mesh. However, how to proceed in case there is a need for seamless modules?

All meshes are comprised of triangles, and those triangles are defined by vertices. Triangles are one-sided, so a normal is needed to define which direction a triangle is facing. However, in a game engine, the normals are not stored per triangle, but per vertex. This allows interpolation between the triangle's vertex normals to create the effect of smooth shading. If there is a mesh that is to be rendered smoothly but also includes hard edges, extra vertices are added to that edge, so that neighboring triangles visually share an edge, but seemingly face

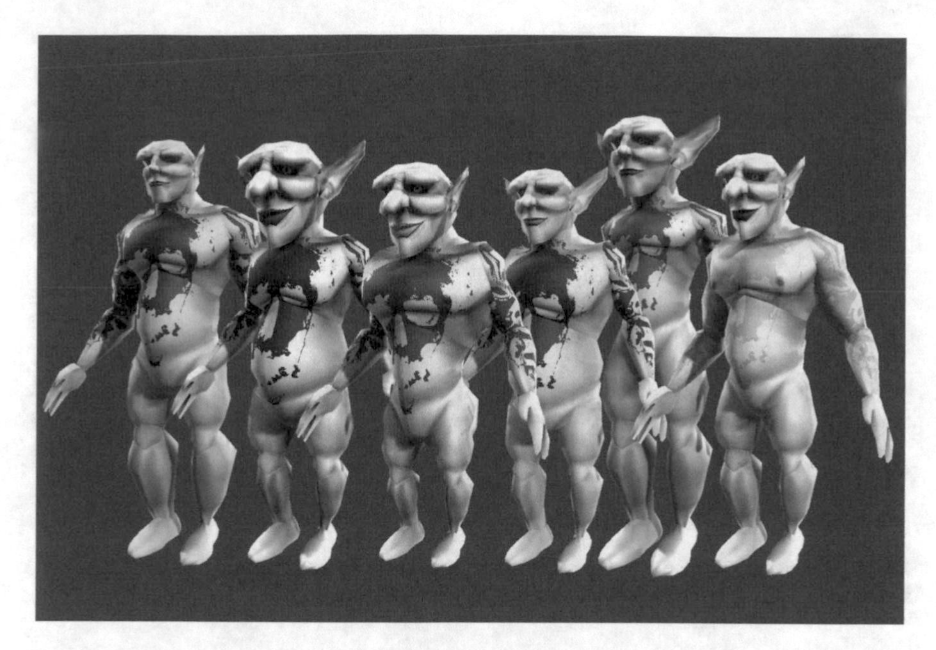

Fig. 4 Examples of creatures generated by the *Invocation Toy* (Gitahy Oliveira 2016), using blend shapes, bone scaling and gradient maps

different directions because of the way the gradients are calculated. That means that on a given model there can exist more than one vertex in a position where in theory only one is visible. Vertex duplication is also used for creating disjoint UV map areas.

This means that if two separate meshes are joined together and need to be seamlessly connected, not only the triangles must line up (i.e., the edge's vertices are in the exact same position), but also the normals must create a smooth gradient between these neighboring triangles. However, since a vertex in a given position might actually be duplicated for shading or UV mapping purposes, how to adjust all of that automatically while not breaking anything (Fig. 5)?

In *Impossible Creatures* or *Spore*, intermediate ways of representing the surfaces were used before generating the final triangle meshes. However, what if working with typical vertex meshes is mandatory? Automatic methods include many requirements, namely making sure that UVs and hard edges are kept intact and, as these characters are aimed at real-time applications, being as little CPU-intensive as possible at runtime. Complete automation can be traded for greater manual control and a simpler method, however: By manually tagging the vertices in both edges, vertices can be translated during runtime from one object to the equivalent position on the other one, and normals can be copied from the host object to the one that was latching into it. We can refer to this as *Vertex Tagging*.

A *Vertex Tag* is a sphere that fetches all vertices that might exist within its radius in a given mesh. Thus, it is possible, outside of runtime, to cache all the vertex

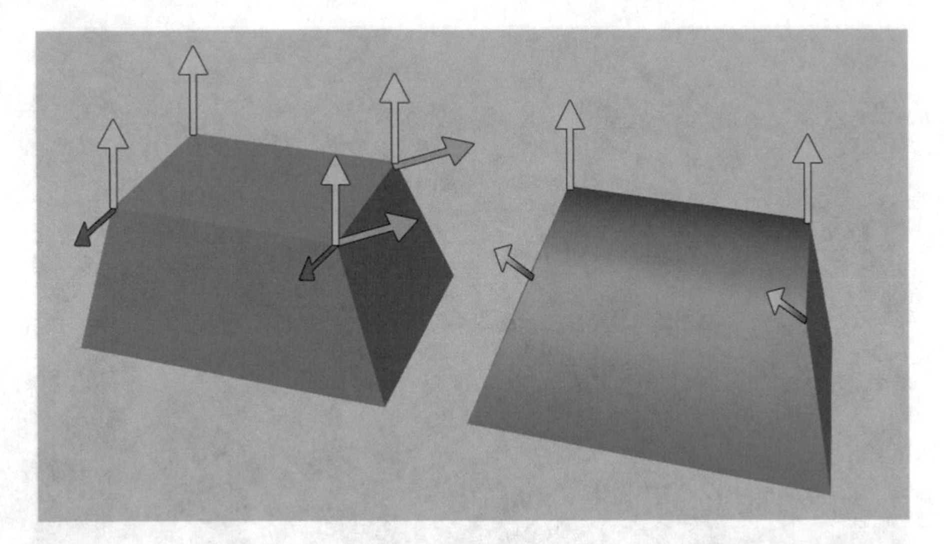

Fig. 5 On the object in the *left*, there is flat shading because of the diverging normals. On the one in the *right*, there is a single normal resulting in a smooth gradient

In Unity3D, whenever any of the array getters in a mesh are used (e.g. to fetch its vertices, normals etc.), a duplicate of the data is created, so make sure to cache the array data before iterating.

```
public void GetVertices() {
  VertexIndexes = new List<int>();
  Vector3[] vertices = GetTargetMesh().vertices;
  Transform trans = GetTargetTransform();

  for (int i = 0; i < vertices.Length; i++) {
      Vector3 worldPos = trans.TransformPoint(vertices[i]);
      if (Vector3.Distance(worldPos, transform.position) < Radius) {
         VertexIndexes.Add(i);
      }
  }
}
```

Listing 2 Using an object's *transform* in Unity to get all vertex indices within a sphere centered in this transform's position, within a given radius

indices of what can be visually classified as "*a vertex in the edge*": a translation between what we refer to when we think of a vertex and how the game engine represents it internally, removing the worry if this "vertex" is actually a group of vertices or not (Listing 2).

A structure to control a group of tags is required. A *Socket* is comprised of several vertex tags, and it defines where parts should "connect" in space. That way, sockets can be used to position the modules themselves (in a way that they properly align) and then can command all the tags to join the proper vertices. In a simpler but still effective solution, joining acts on the tagged vertices only (so the further apart

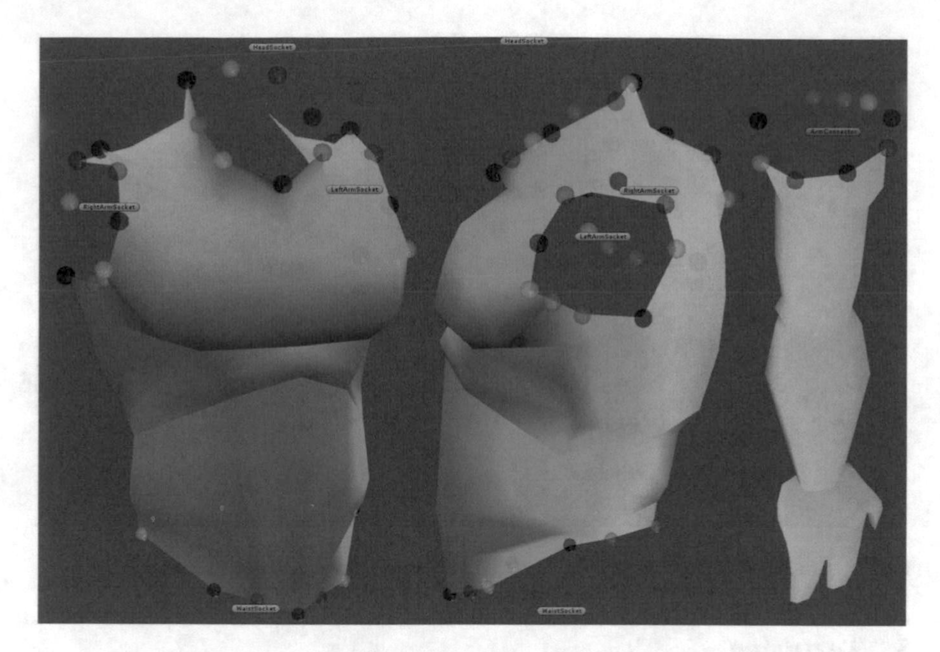

Fig. 6 Views of a model's torso and arms rigged with vertex tags

the objects are, the more deformation it causes), but it would even be possible to use methods as Verlet integration to smoothly "drag" the neighboring vertices along. If the modules are properly authored, however, that will most likely not be needed. If the deformation is too much and cannot be avoided by adjusting the asset creation pipeline, maybe a more complex solution is necessary, like Laplacian surfaces (Huang 2007) or metaballs (Fig. 6).

One advantage of this Socket system over other methods is that it can be improved by creating rules that allow sockets with different amounts of vertex tags to connect: If there are more vertex tags on the host object, the host object might be forced to change, or all the extra vertex tags of the guest object can be merged to the same tag in the host object. Obviously, the best thing is trying to keep the amount either equal on both sides, or very close to that, but it is still interesting to have the ability of having an arm growing out of what should be a neck.

Unlike blend shapes, the Sockets allow a greater freedom in designing different character "modules" that join seamlessly, since they are not limited to small modifications on a single base mesh. However, this freedom comes with a price: If different body parts need different skeletons for animation, bones also need to be merged together and animations have to be merged accordingly (which, again, might depend on more advanced IK solutions). However, if the same rig can be used for all possibilities, this technique paves the way to having an immense amount of variations by combining the previous techniques together (Fig. 7).

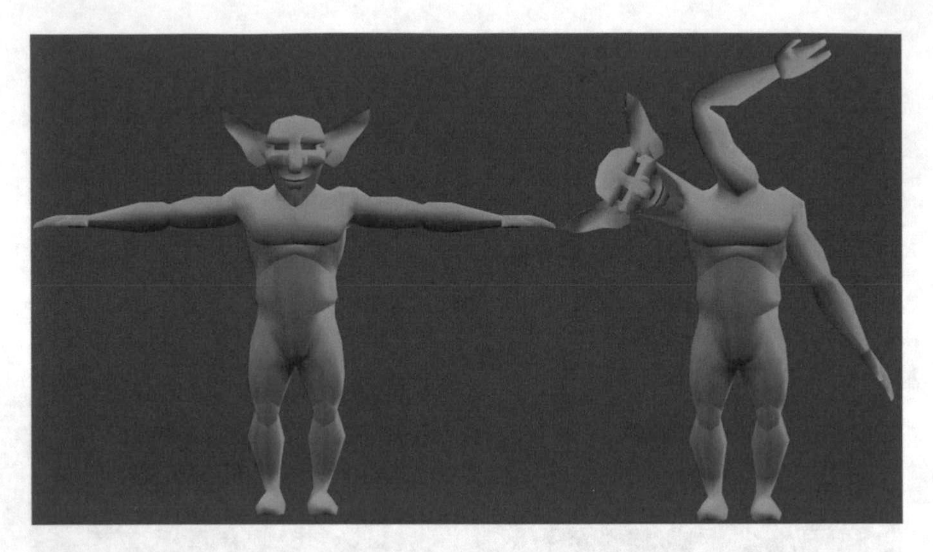

Fig. 7 A model with body parts reorganized via Sockets to join modules seamlessly

5 Texturing

Although silhouettes are very important, color is a huge part of a character. It can be used to indicate its personality, its damage type, its faction and whether it is hostile or not, all at a glance. Obviously, when dealing with procedural characters, it is another dimension of variation.

In older games, render engines depended on color palettes to create variation within the memory budget. Over time, several techniques appeared, like color cycling, where a palette's colors would be cycled to create complex-looking fluid animations (Huckaby 2005). However, the one that got immortalized in games as part of the language of the medium itself was palette switching characters: Mario could have a brother that was fond of green apparel, *Diablo II* could make players instantly realize that Rakanishu was different from the average imp, and *Mortal Kombat*'s rainbow of ninjas could share most frames while looking widely different, still leaving room for all the shades blood.

Currently, however, hardware has gigabytes of memory and compression usually tries to preserve as much as possible the gamut in the original true color textures. That means literal palette swapping is no longer commonplace—but the concept is still widely used and implemented in different ways.

There are several ways of representing a color: RGB (red, green, blue), HSL (hue, saturation and lightness), HSV (hue, saturation, value), etc. However, in all of them, a channel is represented from "none" to "full value." Whatever the final representation is, it is still a group of [0,1] intervals that can be used as a base.

The simplest way to achieve color tinting is multiplying a pixel's color value by the tint color. However, white pixels will become 100% the tinted color, so unless

```
float grayscale = tex2D(_MainTex, IN.uv_MainTex).r;
float3 colored  = tex2D(_Gradient, float2(grayscale, 0.5)).rgb;
```

This snippet roughly translates to: "read the red channel of the main diffuse texture into a float using the UV coordinates from the input, then use that value as an UV coordinate to read out the ramp texture and get the final color". This is possible because just like a color channel, UV coordinates are also a normalized value.

Listing 3 A snippet of a shader that gets a texture's grayscale value and uses it to sample another texture that contains a gradient, in order to generate better color tinting

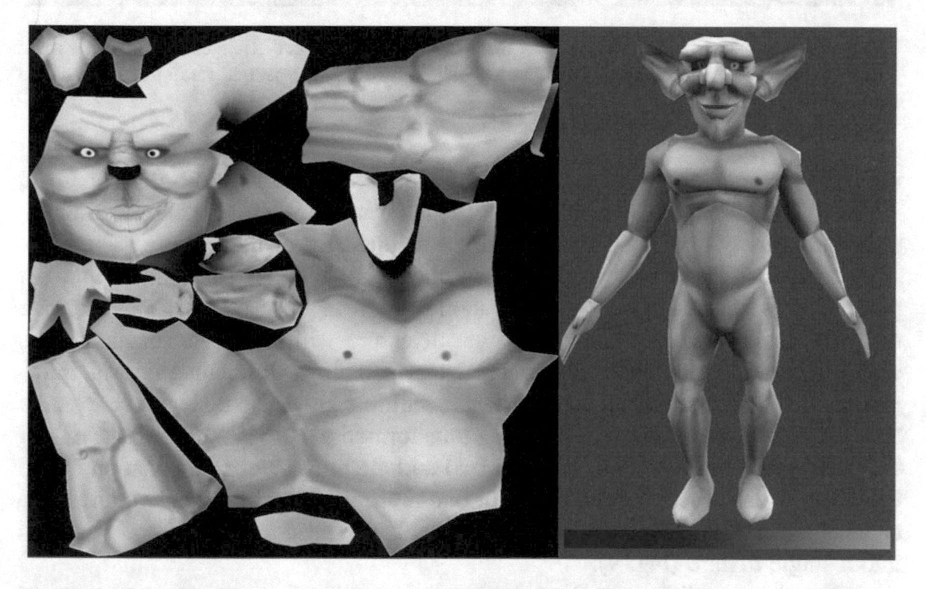

Fig. 8 Applying the blue ramp texture with the grayscale diffuse map, we get the final result without losing the whites

the character is made of metal (metallic object highlights have the same color as the object's color), the result will not look very good.

Gradient mapping is an alternative technique. A gradient is sampled based on a value from 0 to 1. The sampled color is used to color the character. This is very common in authoring tools like Photoshop (via adjustment layers) and can be implemented by a simple shader for real-time use by sampling a texture with the gradient (commonly referred to as a "ramp texture") with the value from another texture (Listing 3; Fig. 8).

If the texture has a very limited amount of grayscale values, the ramp texture can be created to be analogous to a palette if there is no texture filtering (i.e., a texture that would be displayed with a "pixelated" look); otherwise, there might be sampling artifacts. However, for greater color depths and textures that need filtering,

Fig. 9 Two ramps combined via different shader techniques. The red channel is used as the main diffuse texture; the green channel is used as a "palette" that uses the second ramp texture to create the contrasting areas

ramps with a smaller amount of colors may yield better results. With this approach, white highlights can be kept, but that is usually not enough: Monochromatic palettes can be very bland and are generally not suited for characters.

A solution for adding complementary colors is having mask textures that tint parts of the diffuse texture by multiplication or by using other ramps. This is where different channels can be used: The typical texture will have four channels: red, green, blue and alpha, and each channel can contain a different group of information. If shader complexity is not an issue, multiple grayscale textures can even be "packed" into a single channel by doing transformations in UV coordinates, e.g., if the mask exists between (0,0) and (0.5,0.5) and the grayscale for a given part from (0.5,0) to (1,0.5), there is even room for one more mask and grayscale in the same channel (Sosa 2015). UV transformations can also be used to atlas several gradients into a single texture (Fig. 9).

For modern rendering pipelines however, just diffuse textures are not enough: There is the need for normal maps, occlusion maps, etc. *Spore* used a particle paint system, where "smart particles" would be scripted to move in certain ways around the surface of the mesh, while painting at the same time both diffuse and normal maps (DeBry et al. 2007).

Texture memory footprint and shader complexity have to be weighted out by several factors, namely how many characters need to be rendered at a given time. All the regular guidelines apply (for example, atlasing several textures into a single one and attempting to render as many elements as possible in a single batch). Complex multi-texture methods can also be used and then baked down to a single texture that contains the final look for the character if memory is an issue.

There are quite a few techniques for generating procedural color palettes in runtime, but they are also possible to pre-generate, either by hand or by automatically extracting them from concept art or other characters. All of this can be fed into a generator for gradients and used to create countless texture ramps (Listing 4).

Unity3D has a handy Gradient class, which is also serializable. This allows generating gradients at runtime as well, with the possibility of randomizing them. After a gradient is generated, it simply needs to be baked into a ramp texture:

```
private Texture2D GetTextureFromGradient(Gradient gradient) {
  Texture2D tex = new Texture2D((int)FinalTexSize.x, (int)FinalTexSize.y);
  tex.filterMode = FilterMode.Bilinear;

  Color32[] rampColors = new Color32[(int)FinalTexSize.x];
  float increment = 1 / FinalTexSize.x;

  for (int i = 0; i < FinalTexSize.x; i++) {
      rampColors[i] = gradient.Evaluate(increment * i);
  }

  tex.SetPixels32(rampColors);
  tex.Apply();

  return tex;
}
```

Listing 4 Creating a ramp texture in Unity based on a gradient by sampling it with regular intervals

6 Conclusion

In this chapter, we have covered methods for modular character creation in 2D and 3D, with a small recapitulation of fundamental knowledge about how meshes are represented and lit in games and real-time applications. *Spore* is still one of the best references for fully procedural characters, and fortunately, there are several publications and blog posts about the methods used by it.

No generator will have a good result if it is not designed with purpose and art direction in mind, always considering the best practices of character design. Combining just a few elements can lead to a great number of variations, but the players will eventually notice where all of them come from, so creating meaningful variation is something to worry about: Are the characters cannon fodder in a zombie shooter, or are they the very center of the experience? How many of them need to be rendered at the same time? Is it okay for them to move around in a silly way? This all weights in what techniques need to be used.

It is rare to see a major game that truly relies on procedural characters because art, design and technology need to be at their finest. That takes time and budget, which calls for guaranteed financial results, but even with its presence in the mainstream in the recent years, procedural generation is still very much an experimental beast that has no common language that is fully understood between developers and players, as pointed out by Michael Cook in his article *Alien Languages* (Cook, Michael, 2016). However, smaller independent studios constantly explore it with simpler aesthetics and, bit by bit, push the boundaries of the medium. One would dare say that building a game around procedurally generated characters has a terrible cost–benefit relation in a commercial game, but it is too enticing a challenge to let go; it is about developers indulging in a different kind of wanderlust.

Procedural generation is ultimately about iteration, about "sculpting" algorithms until they get to a point where they have varied results but, most importantly, achieve what is needed of them—and that is especially true when it comes to bringing life into virtual beings.

7 Future Work

Animation of procedural characters is a theme worthy of its own book. When characters can share skeletons and animations, the workflow is the same as for regular characters. However, if they have shapes different enough to require multiple or also modular rigs, there is little support out of the box from game engines. Solutions will most likely involve inverse kinematic systems with varying degrees of complexity and, even though there are several publications regarding IK, character gait, motion generation and animation retargeting are still very active research areas, ranging from procedural animation of multi-legged creatures (Karim, Gaudin, Meyer, Buendia, & Bouakaz, 2013) to adapting motion capture data to any possible skeleton (Bereznyak, Alexander, 2016).

Future work includes making sure that Vertex Tags and Sockets support properly joining bones into full skeletons and making sure that the vertices on the edges are properly weighted for skinned animation. With this comes the actual challenge: How to make sure that wildly different modules animate properly in conjunction (e.g., the imp pictured previously, except with a slithering, snake-like lower body)? Unity3D has a full-fledged animation system, but it is currently not possible to alter animation files to merge different parts together during runtime, so maybe it would be necessary to either implement a robust IK system or a custom animation playback system specialized in modular parts.

While a lot of the technology facet was covered in this chapter, what are some of the art pipeline challenges? How to make sure wildly different variations and modules fit together? When using normal maps, is it still possible to make surfaces seamless? How to design modules that can be wildly altered by approaches like blend shapes or similar to Spore's *Rigblocks*?

The recent *No Man's Sky* could be considered *Spore*'s spiritual successor, also an excellent reference for modern rendering needs, and information about it will be published in the coming years as well. One of the most interesting (and unsung) aspects of the game is that even the creature sounds are procedural and based on their size and shape. While procedural music is somewhat commonplace, there is a lot of room for experimentation in procedural sound effect generation.

Finally, how can one measure the impact of a randomly generated character? How long until the player realizes all variants stem from a smaller group of elements—and how to make the best out of them in a meaningful way, gameplay-wise? What are some other interesting generational approaches that do not rely on modular parts? All of these topics need to be studied further and experimented with. As writers say: "every character in a story thinks they're the main character"—and it is up to our assets and algorithms to make them so.

References

Bereznyak, Alexander. (2016). IK Rig: Procedural Pose Animation. Presented at the Game Developers Conference. Retrieved from http://schedule.gdconf.com/session/ik-rig-procedural-pose-animation

Boneloaf. (2015). *Gang Beasts*. Retrieved from http://gangbeasts.game/

Choy, L., Ingram, R., Quigley, O., Sharp, B., & Willmott, A. (2007). Rigblocks: Player-deformable Objects. In *ACM SIGGRAPH 2007 Sketches*. New York, NY, USA: ACM. 10.1145/1278780.1278880

Cook, Michael. (2015). PROCJAM. Retrieved September 22, 2016, from http://www.procjam.com/

Cook, Michael. (2016, August 18). Alien Languages: How We Talk About Procedural Generation. Retrieved September 23, 2016, from http://www.gamesbyangelina.org/2016/08/procedurallanguage/

DeBry, D. (grue), Goffin, H., Hecker, C., Quigley, O., Shodhan, S., & Willmott, A. (2007). Player-driven Procedural Texturing. In *ACM SIGGRAPH 2007 Sketches*. New York, NY, USA: ACM. 10.1145/1278780.1278878

Dunlop, Drew. (2002, November 18). Impossible Development Diary. Retrieved from http://www.ign.com/articles/2002/11/18/impossible-development-diary

Gitahy Oliveira, Yanko. (2015). *X, a game of YZ*. Retrieved from https://itch.io/jam/procjam2015/rate/43625

Gitahy Oliveira, Yanko. (2016). *Bestiarium Invocation Toy*. Retrieved from https://yanko.itch.io/invocation-toy

Haxby, J. V., Hoffman, E. A., & Gobbini, M. I. (2002). Human neural systems for face recognition and social communication. *Biological Psychiatry*, *51*(1), 59–67. 10.1016/S0006-3223(01)01330-0

Hecker, Chris. (2014, August 7). Liner notes for Spore. Retrieved from http://chrishecker.com/My_liner_notes_for_spore

Hecker, C., Raabe, B., Enslow, R. W., DeWeese, J., Maynard, J., & van Prooijen, K. (2008). Real-time Motion Retargeting to Highly Varied User-created Morphologies. *ACM Trans. Graph.*, *27*(3), 27:1–27:11. 10.1145/1360612.1360626

Hello Games. (2016). No Man's Sky. Retrieved September 22, 2016, from http://www.no-mans-sky.com/

Huang, X. (2007). Optimal boundaries for Poisson mesh merging. Retrieved from http://www.cs.jhu.edu/~misha/ReadingSeminar/Papers/Huang07b.pdf

Huckaby, Joseph. (2005). Old School Color Cycling with HTML5. Retrieved from http://www.effectgames.com/effect/article-Old_School_Color_Cycling_with_HTML5.html

Inagaki, K., & Hatano, G. (2006). Young Children's Conception of the Biological World. *Current Directions in Psychological Science*, *15*(4), 177–181. 10.1111/j.1467-8721.2006.00431.x

Karim, A. A., Gaudin, T., Meyer, A., Buendia, A., & Bouakaz, S. (2013). Procedural locomotion of multilegged characters in dynamic environments. *Computer Animation and Virtual Worlds*, *24*(1), 3–15. 10.1002/cav.1467

Maxis. (2008). Spore. Retrieved September 22, 2016, from http://www.spore.com/

Mitchell, J., Francke, M., & Eng, D. (2007). Illustrative Rendering in Team Fortress 2. In *Proceedings of the 5th International Symposium on Non-photorealistic Animation and Rendering* (pp. 71–76). New York, NY, USA: ACM. 10.1145/1274871.1274883

Pope, Lucas. (2012, November 25). "Papers, Please" devlog. *Papers, Please - A Dystopian Document Thriller*. Retrieved from https://forums.tigsource.com/index.php?topic=29750.20
Pope, Lucas. (2013a). Papers, Please. Retrieved September 22, 2016, from http://papersplea.se/
Pope, Lucas. (2013b). *Papers, Please - Trailer*. Retrieved from https://www.youtube.com/watch?v=_QP5X6fcukM
Relic Games. (2003). Impossible Creatures. Retrieved September 22, 2016, from http://store.steampowered.com/app/324680/
Rieth, C. A., Lee, K., Lui, J., Tian, J., & Huber, D. E. (2011). Faces in the Mist: Illusory Face and Letter Detection. *I-Perception*, *2*(5), 458–476. 10.1068/i0421
Schwind, V. (2015). Historical, Cultural, and Aesthetic Aspects of the Uncanny Valley. In C. Misselhorn (Ed.), *Collective Agency and Cooperation in Natural and Artificial Systems* (pp. 81–107). Springer International Publishing. Retrieved from http://link.springer.com/chapter/10.1007/978-3-319-15515-9_5
Sokpop. (2016). Sokpop. Retrieved from http://sokpop.co/
Sosa, Jesse. (2015, January 25). Modular Character Workflow. Retrieved from http://wiki.polycount.com/wiki/SkankerzeroModularCharacterSystem

Author Biography

Yanko Oliveira started out as an indie developer during his computer science studies at the Federal University of Rio de Janeiro, releasing mobile and Leap Motion games. He then went on to work at Aquiris Game Studio, Brazil's top game company, being a developer in Ballistic, Cartoon Network Superstar Soccer, and Horizon Chase, the latter which won several accolades and was released as an AppStore Editor's Choice.

Oliveira moved to Hamburg, Germany, to work at GoodGame Studios, and is currently lead developer at Bigpoint Games. After treating his wife's arachnophobia with his master's thesis game, he now focuses his free time on experimental prototypes, mostly exploring procedural generation and VR, still releasing his indie ventures as deVoid Games.

You can follow the development of his personal projects at http://yankooliveira.com, http://yanko.itch.io and Twitter, @yankooliveira.

Procedural Terrain Generation. A Case Study from the Game Industry

Jakob Schaal

Abstract This chapter shows how PCG can be used for landscape generation in games. A very brief introduction to value noise generation is provided. Any noise generator capable of generating cloud pictures can generate similar results with the new algorithm, for example the well-known Perlin noise or its derivation, the simplex noise. We then provide both basic algorithms and practical hints for generating different types of terrain. A new algorithm is presented which generates landscapes with islands of different size and levitation. This algorithm has been created for an industry game project to increase the variety of islands in an explorer game. We show in detail how noise-based images generate a 3D-Terrain, how this terrain can be manipulated so it looks realistic and how the landscape can be textured. The techniques used are not specific to any game engine—they can be implemented in any 3D engine capable of creating custom meshes at runtime.

1 Introduction

When generating terrains with a PCG Algorithm, a semi-endless amount of different game levels can be created. This is an advantage to hand crafted, static terrains, because the player of a video game never runs out of new content. To make sure that the different resulting terrains "feel" different to the player, the algorithm must be capable of creating different variations of the terrain. Some methods to do this are provided in this chapter.

There are different approaches when generating terrain with PCG. The most common one is to generate a noise function as illustrated in Fig. 1. Such a noise function can then be used as a height map for a terrain, meaning that higher values translate to higher points in the generated terrain. When illustrating a noise function as an image, it is common to use brighter pixel colors for higher values and vice versa.

J. Schaal (✉)
Stuttgart Media University (HdM), Stuttgart, Germany
e-mail: jakob.schaal@hotmail.com

© Springer International Publishing AG 2017

O. Korn and N. Lee (eds.), *Game Dynamics*,
DOI 10.1007/978-3-319-53088-8_8

Fig. 1 An example of a Value Noise

Different noise generators are used today which normally lead to similar results. The differences are commonly limited to computational costs as well as artifacts. Some methods are better suited for higher dimensions, like the Simplex Noise (Gustavson 2005). The following chapter utilizes a Value Noise (2016), which is a comparably fast and easy to understand approach for two dimensions with low artifacts. However, other methods can be used as well, for example the notorious Perlin Noise (Perlin 1985). The game Minecraft (Mojang Synergies 2009) utilizes a modified Perlin Noise to generate a semi-endless terrain which the player is able to traverse.

When generating a 3D Level, different approaches are possible, too, depending on the game rules. For example, the games *Diablo III* (Blizzard Entertainment, Inc. 2012) and *The binding of Isaac* (McMillen and Himsl 2011) shuffle predefined, handcrafted level parts together to generate a complete level. Terrain generation can be applied as a part of a bigger PCG algorithm. For example, the game *No Man's Sky* (Hello Games 2016) creates a whole universe with different solar systems and planets. Theses planets also use terrain generation for their surface. Generating content by algorithms instead by hand is no new discovery. The game *Elite* (Braben and Bell 1984) already made heavy use of PCG Algorithms in 1984 to generate galaxies and solar systems. PCG is not only limited to generating terrains. It can also be used to generate racing tracks (Botta et al. 2012) or buildings (Müller et al. 2006). Completely different approaches can be used to generate terrain like evolutionary algorithms (Raffe et al. 2012; Ong et al. 2005) or cellular automats (Johnson et al. 2010). A good overview of the different kinds of PCG is given by Hendrikx et al. (2013).

This chapter explains how PCG methods can be used to generate a 3D terrain in an industry game environment. After providing some technical background, we discuss content generation approaches, like the generation of volcanic islands. The following section provides a use case for the provided algorithm: a future top-down tactical game with the work title "New Horizon Procedural" (NHP). The player is able to control various ships and navigate them through a Caribbean landscape. The player can build harbors at predefined spots, which the algorithm selects. Generating an endless amount of levels with one algorithm was a huge goal when developing NHP. As this chapter is practically oriented, it focuses on the discussion of the algorithm steps.

2 State of the Art

2.1 *Value Noise*

This section gives a short explanation of a Value Noise generation. Different Value Noise algorithms exist that differ in the interpolation method used. It is possible to interpolate via a simple linear interpolation function; however, cubic interpolation methods lead to less artifacts and should therefore be preferred.

In Fig. 2, random *y*-values along fixed positions along the *x*-axis are illustrated. These values are called nodes. To interpolate any value between these nodes, first the neighboring values (the red dots) have to be picked, labeled as *a* and *b*. After that the percentage value of the distance along the *x*-axis from *a* to the interpolated point has to be calculated, whereas the distance from *a* to *b* along the *x*-axis is 100% or 1.0. This percentage value then is applied to a third variable *t*. Thus if $t = 0$ then the *x* value is exactly the *x* value of the point *a*, if $t = 1$ it is the *x* value of the point *b*, if $t = 0.5$ it is in the middle and so on. With these three variables, *a*, *b* and *t*, the lerp (linear interpolation) function can be defined as:

$$\mathrm{lerp}(a, b, t) = a * (1 - t) + b * t$$

Once two additional variables are defined, preA and postB, a cubic interpolation method (Bourke 1999) can be used instead, which leads to less artifacts in the

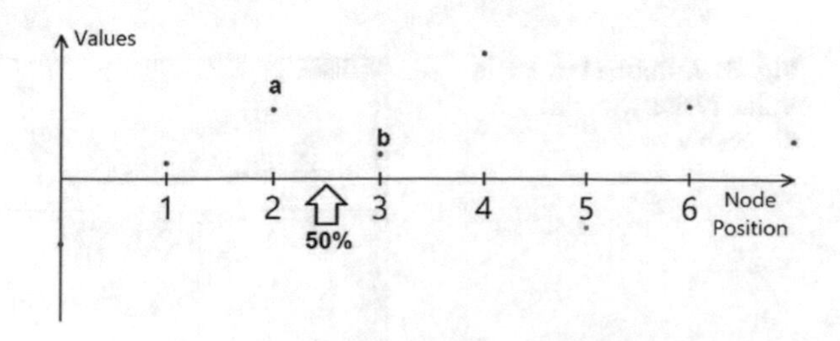

Fig. 2 Nodes are placed with constant distance along the *x*-axis

resulting noise function. preA is defined as the value preceding a, and postB is defined as the value after b. This leads to a cubic interpolation method as follows:

$$\text{cubic}(\text{preA}, a, b, \text{postB}, t) = (\text{postB} - b - \text{preA} + a) * t^3 + (2 * \text{preA} - 2 * a - \text{postB} + b) * t^2 + (b - \text{preA}) * t + a \quad (1)$$

which translates to the following C#-Code (with optimizations):

The calculation of a complete Value Noise is subdivided into several "octaves". Once every point of Fig. 2 is interpolated as shown in Fig. 3, the result is one octave of a one-dimensional Value Noise. A Value Noise has a fixed number of octaves, for example 5. With every octave, the range of possible values for the nodes is decreased (for example halved) and the number of nodes is increased (for example doubled). An example for a second octave is illustrated in Fig. 4. This octave generation is repeated until the desired number of octaves is reached. Once every octave is generated, the values are added and the resulting function is a Value Noise, as seen in Fig. 5.

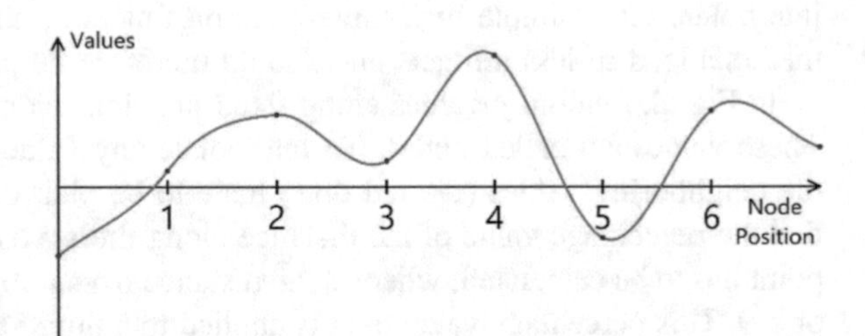

Fig. 3 First octave of a Value Noise

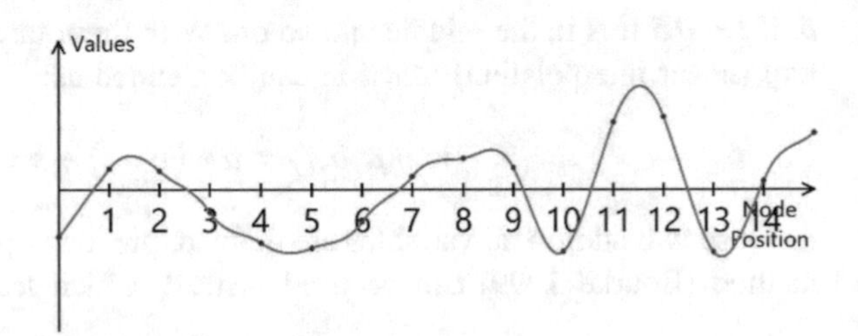

Fig. 4 Second octave of a Value Noise

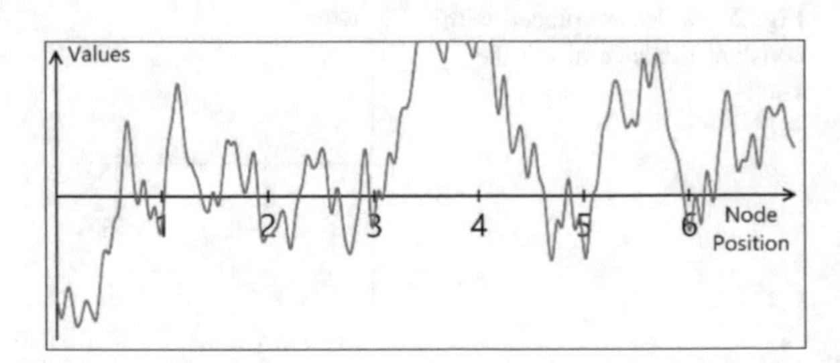

Fig. 5 A finished example Value Noise with one dimension

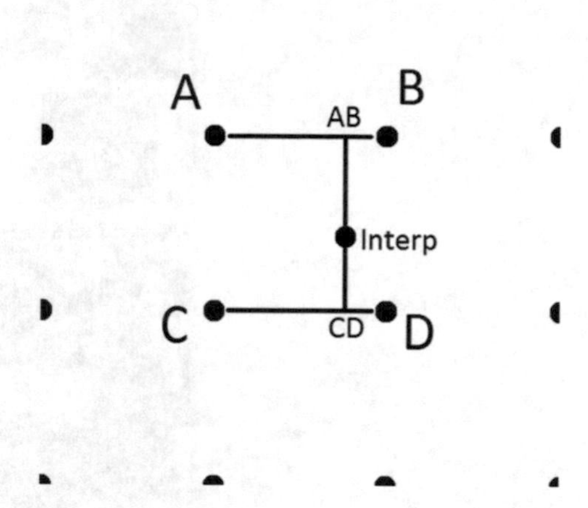

Fig. 6 A grid with bilinear interpolation

Figure 5 shows a one-dimensional Value Noise. However, for a terrain generation at least two dimensions are required to generate a usable height map. To generate a two-dimensional Value Noise, first a grid of nodes is created as illustrated in Fig. 6. In the second step, every pixel of the grid has to be calculated. To achieve this, a bilinear interpolation is performed for every pixel. This means that the four edge nodes of the pixel which is about to be calculated are picked (in Fig. 6 they are called A, B, C, and D). First, the horizontal lines are interpolated as seen in the one-dimensional case and afterward a last interpolation in the vertical dimension is done, which results in the value of the desired pixel. Once every pixel is calculated, the result is the first octave for a two-dimensional Value Noise, as seen in Fig. 1. As explained in the one-dimensional case, other octaves have to be calculated as well, increasing the density of the nodes in the grid and lowering their possible range of values. Once all octaves are calculated, their values are added again and the result is a two-dimensional Value Noise. The resulting pixel image is directly translated to a height map for a 3D terrain. Note that such a terrain does not have any caves because every two-dimensional position only has one height value in the height map.

3 Implementation

3.1 Generating the Terrain

Value Noise can be used to generate islands. These islands are placed inside a 3D game map, which consists of various islands and the sea. To know the exact attributes for a single Value Noise (for example its height and width) the position

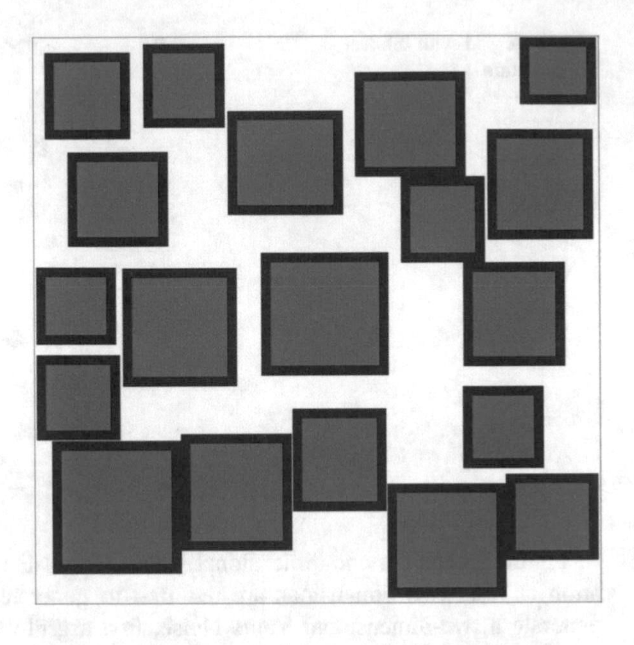

Fig. 7 Island spot placement

for such an island has to be calculated first. This is done via simple rectangle placement with intersection checks. A rectangle with random width and height is placed inside the map, and it is checked if the rectangle intersects with any previously placed rectangle. If so, the rectangle is rebuilt with different height, width and position. If not, the rectangle is used as an island spot. The result of such a rectangle placement is shown in Fig. 7. Every green rectangle is the spot for an island. The red part is an additional "safety distance" to guarantee that the ships of the player are able to navigate between those islands. Tests have shown that if the islands have roughly the same height and width (meaning that they are approximate squares), the result looks better.

It is also possible to multiply the Value Noise with a different random number for every island. This leads to more variation, as some islands are flatter and some feature higher ground.

If a Value Noise is directly used as a height map, then the borders of such an island would get cut of immediately which leads to visually unpleasant results as seen in Fig. 8. To avoid this problem, an additional mask is generated and multiplied with the result of each Value Noise. This mask has three requirements.

1. A big part of its center has to be 1, meaning that the value of the Value Noise is unchanged.
2. The edge of the mask must be 0, ignoring the Value Noise completely.
3. The area between those two extremes has to fade smoothly.

Such a mask can be generated using the following equation:

Fig. 8 If the noise is directly used as a height map for the terrain, the edges do not blend smoothly with the sea

$$\text{mask}(x, y) = \text{clamp}\left(2 * \left(1 - \frac{\left|\frac{w}{2} - x\right|}{\frac{w}{2}}\right)\right) * \text{clamp}\left(2 * \left(1 - \frac{\left|\frac{h}{2} - y\right|}{\frac{h}{2}}\right)\right) \quad (2)$$

with:

- w = width of the mask = width of the Value Noise
- h = height of the mask = height of the Value Noise

and:

$$\text{clamp}(x) = \begin{cases} 0, & |x < 0 \\ 1, & |x > 1 \\ x, & |\text{else} \end{cases} \quad (3)$$

This translates into the following C# Code:

```
private float w;
private float h;



//Constructor etc.


public float mask(float x, float y){
   float wHalf = w / 2;
   float hHalf = h / 2;
   float w1 = clamp(2 * (1 - Math.abs(wHalf - x) / wHalf));
```

```
        float w2 = clamp(2 * (1 - Math.abs(hHalf - x) / hHalf));
        return w1 * w2;
    }



    private float clamp(float x)
    {
        if (x < 0) return 0;
        if (x > 1) return 1;
        return x;
    }
```

A result of such a mask calculation can be seen in Fig. 9. Once the mask is multiplied with the Value Noise, a smooth transition between island and sea is generated as seen in Fig. 10.

3.1.1 Sea Ground

The islands now have an interesting structure. However, the ground of the sea looks drab; it is just flat everywhere and has the exact same height at every point. Interesting structures like sandbanks would lead to a visually more pleasant look.

Fig. 9 Result of a mask calculation

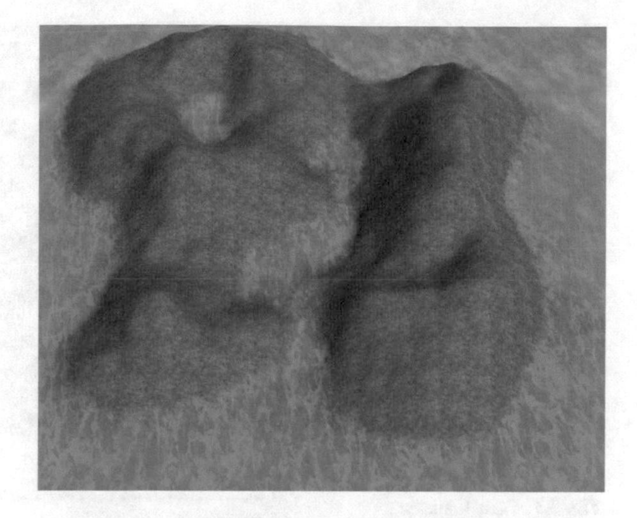

Fig. 10 Once the mask is multiplied with the height map, the edge of the island blends smoothly with the sea

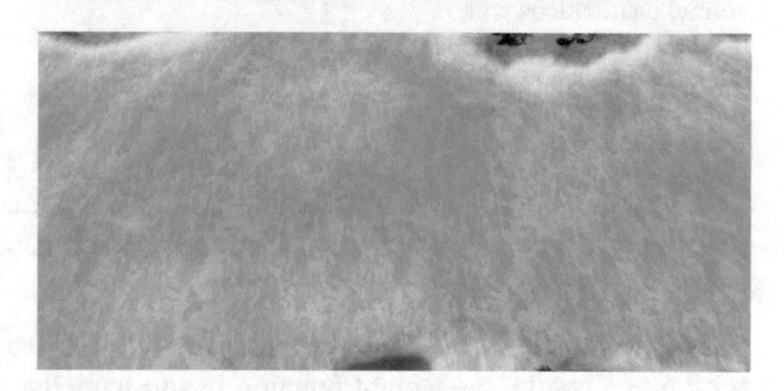

Fig. 11 Another Value Noise is generated to create under water structures

To achieve such a sea ground structure, another noise is generated which has a relatively low maximum height. This noise is added to the height of the whole map instead of just a single island. This means that the noise must have the size of the whole map.

After adding this sea ground noise, structures occur as illustrated in Fig. 11. If these sand banks should not interfere with the gameplay, it is important to set the maximum height of the noise below sea level. This way no sand banks reach out of the water, and only underwater structures are generated. If they would interfere, there is no guarantee that some parts of the map are not completely cut off, making them unreachable to the player's ships.

3.2 *Texturing the Terrain*

The island in Fig. 10 looks monotonous because it only has one texture: grass. To make visually appealing islands, additional textures have to be added. There are different ways to texture a terrain, but we assume that four textures are given: sand, grass, stone and mountains, which all have their highest opacity at different terrain heights. This means that every texture has a weight number, which directly

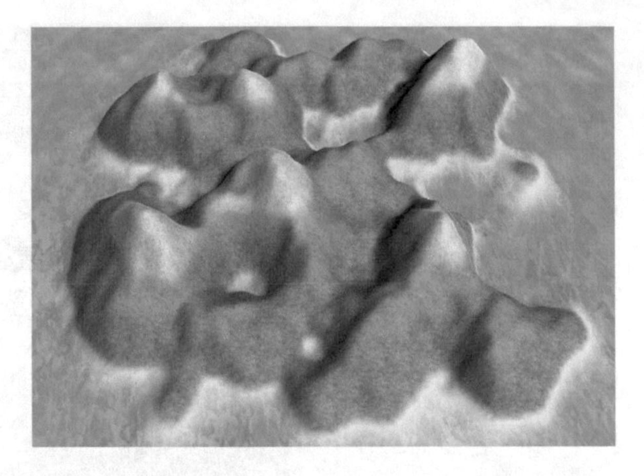

Fig. 12 Texture weights are applied to the terrain

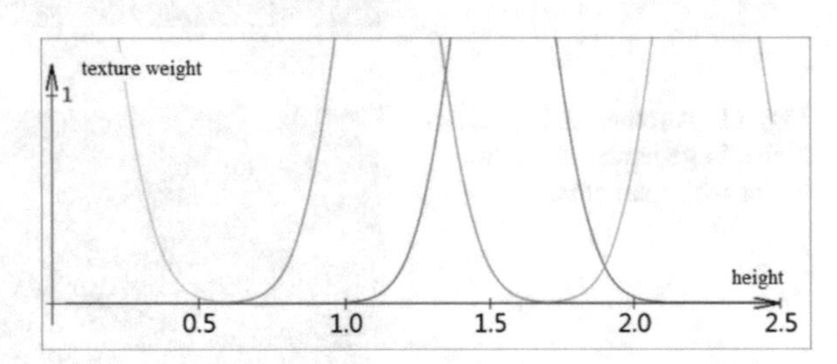

Fig. 13 Four Gaussian normal distributions with different values, representing the weight of the textures: *yellow* = sand, *green* = grass, *brown* = stone, *gray* = mountain

translates to their opacity and their sums are one. To generate the weight of a texture for a given height, the weight function has to have its maximum height at a certain value and drop off when it gets higher or lower. The normal (or Gaussian) distribution fulfills this property (Fig. 12).

The normal distribution is defined as:

$$f(x) = \frac{1}{\sqrt{2\pi}} e^{-\frac{1}{2}\left(\frac{x-\mu}{\sigma}\right)^2} \quad (4)$$

where σ and μ can be freely picked and e is the Euler constant. Assuming that the maximum height is at 2.5, the following values of σ and μ lead to a good result:

Sand: $\sigma = 0.15$ $\mu = 0$
Grass: $\sigma = 0.15$ $\mu = 1.1$
Stone: $\sigma = 0.15$ $\mu = 1.5$
Mountain: $\sigma = 0.15$ $\mu = 2$

This results in the curves illustrated in Fig. 13. However, the sum of these curves generally does not equal one. To avoid this, a normalization step is required: the

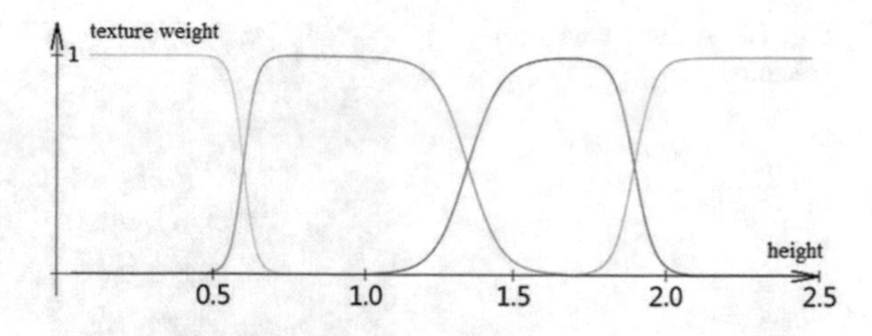

Fig. 14 Same as Fig. 11 but the values are normalized. Their sums are always 1, for every height

Fig. 15 An island with volcanos

sum for every height of every texture is calculated and then every curve value is divided by this sum, leading to a total sum of 1 for every height, as illustrated in Fig. 14. The result is no sudden change in texture but a smooth transition. The result after applying these values to the island is provided in Fig. 12.

3.3 Generating Landmarks

Additional noise transformations can lead to additional interesting phenomena when generating the terrain structure. One such transformation is easy to implement and leads to good-looking areas in the environment: the generation of volcanos. However, to avoid reappearing volcanos not every island should be selected for volcano placement. This leads to a bigger surprise effect on the user once he or she finally discovers a volcano island.

To generate a volcano island, first a maximum height is picked. Every value above this height is flipped. Formally, this leads to the following transformation equation:

$$\text{volcano}(x, y) = \begin{cases} x, & |x < y \\ 2y - x, & |\text{else} \end{cases} \tag{5}$$

where y is the maximum height of the island. Every value of the Value Noise is transformed by this volcano function. This leads to effects as shown in Fig. 15.

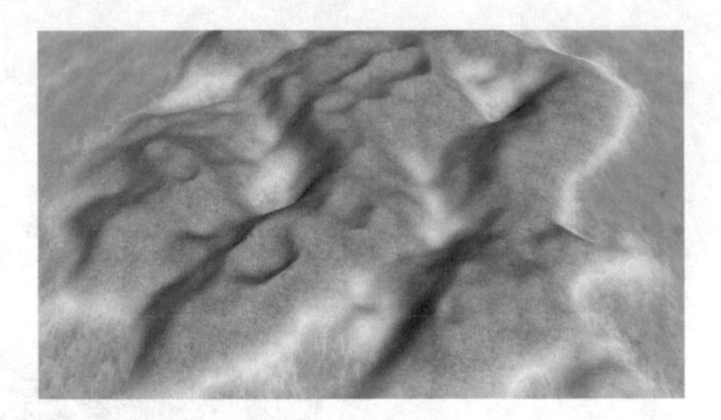

Fig. 16 A flat island with volcanos

In this case, the volcano was generated on an island with high mountains. One might think that the volcano generation transformation should only take place on islands with such high mountains. However, this is a misconception: when the volcano transformation is applied to flat islands, visually appealing effects occur as well, as shown in Fig. 16. This is also realistic because small flat islands normally are created in areas with high volcano activity ("Hotspots"). The flat islands can be interpreted as such "Hotspot islands".

It is possible to modify the volcano function so that if a volcano is generated, the position is saved. This position can then get a different texture (like a lava texture) and can be used to simulate volcano eruptions.

Further transformations of the terrain can lead to additional interesting landmarks. A good approach is to combine the original noise with some other mask, like a Voronoi diagram (Smelik et al. 2008).

3.3.1 Harbor Spots

Harbor spots have to be placed by the algorithm so that the player is able to build a harbor at these locations. These spots are predefined locations generated by the algorithm, which have to be at the edge of an island. Finding a location which is below a certain height is not enough because the noise algorithm is able to generate lakes inside of an island. However, harbor spots must not be generated at these lakes, as seen in Fig. 17.

To find the edge of an island, a flood fill (Flood fill 2016) algorithm is used. First, the height map of the whole 3D Game level is generated. Then another two-dimensional array (state array) with the same size of the height map is created. Every index of the state array can hold values for *sea*, *island* and *edge*. It is initialized with the state *island*. This means that, if the algorithm would terminate at this point, it would assume that the whole map is one enormous island. This obviously is false. The indices at the border of the two-dimensional state array are put into a list (checklist) and their values are set to *sea*, which is the border of the world map. This is always true because every island has a safety distance, as seen in Fig. 7. As a result, islands are never placed at the edge of the map.

Fig. 17 If the harbor placement would only consider height, harbors might be placed at the *red* areas, which are not reachable by ships

Next, the flood fill algorithm is performed as follows:

1. Take the first element from the checklist and remove it.
2. Set its state in the state array to *sea*.
3. Check each neighbor: if it has the state *island* and its height is below sea level, add the neighbor to the checklist.
4. If there are still elements in the checklist, go back to 1.
5. Terminate flood fill.

When picking the neighbors, two different approaches are valid. Either the horizontal and vertical neighbors are considered as neighbors, or the diagonals are considered as well. As the state array can later be used for path finding algorithms, like A* or JPS, the neighbor method which applies to valid movements of the ships should be picked.

After the flood fill algorithm, every sea and island pixel in the height map is marked as such. Another iteration through the whole state array is performed. If an *island* state is found and one or more neighbors are set to *sea*, the value of the index is changed to *edge* instead. These edge indices are put into another list (edge list). To place harbor spots, the following algorithm is performed:

1. Take a random element from the edge list.
2. Place a harbor spot at that location.
3. Remove the element from the edge list as well as all other elements that are close to it.
4. If there are still elements in the edge list and not enough harbors are placed yet, go back to 1.
5. Terminate harbor placement.

This makes sure harbors are not placed too close to each other. Figure 18 shows the result of such a harbor placement. To guarantee that a harbor is not placed at a

Fig. 18 Three harbor spots placed on an island, which also features a volcano in the *middle*

Fig. 19 A result of the first vegetation placement approach. The trees are equally distributed on the island

steep clive which is next to the sea, it is recommended to flatten the terrain once the harbor is placed.

3.4 Generating Vegetation

Other than height, the islands currently have no other three-dimensional structure. To improve this, trees and other vegetation are created and spread on these islands. Vegetation should not be created below grass level of the texture (trees normally do not grow on the sand of beaches) and not above stone level (generally few trees grow on high mountains). This section explains two different approaches for finding valid spots for trees.

The first approach is trivial. Calculate a random position on the island and check its height. If it is above sand and below mountain level, place the tree on the height of the terrain. This leads to a vegetation distribution as seen in Fig. 19. While the

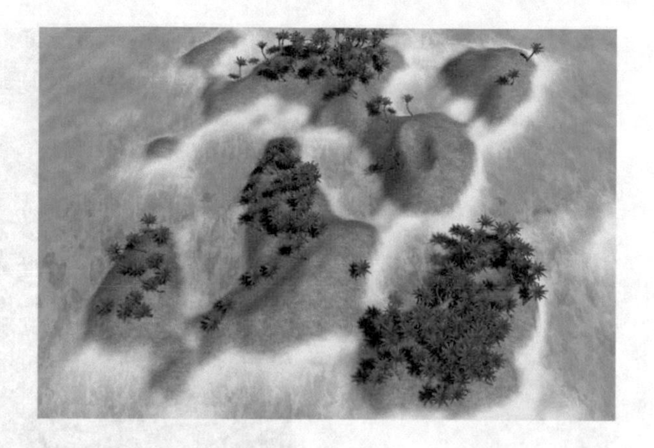

Fig. 20 Result of the second vegetation placement approach. A Value Noise was used to manipulate the distribution and density at different areas of the island

result looks better than before, it still does not look very realistic. The trees are just randomly placed around the island.

To improve this, a second approach was developed: it first generates an additional noise (for example another Value Noise) and maps every position of the terrain to a position in the newly generated noise. This noise represents a probability distribution when placing a tree. Once a random position is calculated as in the first approach, another random value is calculated. If this random value is above the value in the new noise, the tree is placed, else the position is discarded. This leads to results as illustrated in Fig. 20. This is more realistic because it generates both forests and places where almost no trees grow. However, even in places where few trees grow, there still is a probability for a tree—just like in nature.

In the second approach, the noise should not be used directly. Instead, its values should be normalized, meaning that the highest value should become 1, the lowest 0 and every other value should be distributed between 0 and 1. The resulting, normalized noise should then be raised to a power of X. This results in denser forests. If the power is 1, the result is similar to approach one. If it is higher, like $X = 8$, the results are similar to Fig. 20. If it is higher, the forests get even more dense. A Value Noise which is raised to the power of 8 can be seen in Fig. 21.

It is also possible to place other vegetation than trees, like smaller bushes. These could use the same Value Noise but with another, lower value of X. This would lead to bushes that are placed not just inside of forests but also around them. It can of course also be modified to place larger trees inside of forests, with a higher X value. The vegetation itself could also be procedurally generated, for example with SpeedTreeRT (Interactive Data Visualization 2009). Other props like procedural stones can be placed as well (Dart et al. 2011).

Fig. 21 Value noise of Fig. 1 was raised to the power of eight, making the resulting forests denser

4 Conclusion

A new algorithm is presented which generates landscapes with islands of different size and levitation. This algorithm has been created for an industry game project to increase the variety of islands in an explorer game. We show in detail how noise-based images generate a 3D-Terrain, how this terrain can be manipulated so it looks realistic and how the landscape can be textured. The techniques used are not specific to any game engine—they can be implemented in any 3D engine capable of creating custom meshes at runtime.

This chapter showed how PCG can be used for landscape generation in games. Based on an industrial game project, we showed a way to generate a Caribbean island landscape. We discussed the use of different Value Noises to procedurally generate islands of different size and levitation, place the vegetation and add a structure for the sea ground. Additionally, an approach for finding the islands' edges was provided. The described algorithms can be used in gaming productions but is not limited to this application.

In the future, different masks for blending the island with the sea can be researched. In overall, the rectangular mask works—however, if there are enough islands visible, their rectangular shape becomes apparent.

References

Blizzard Entertainment, Inc. (2012). *Diablo III*. Blizzard Entertainment, Inc.

Botta, M., Gautieri, V., Loiacono, D., & Lanzi, P. L. (2012). Evolving the optimal racing line in a high-end racing game. In *2012 IEEE Conference on Computational Intelligence and Games (CIG)* (pp. 108–115). 10.1109/CIG.2012.6374145

Bourke, P. (1999, December). Interpolation methods. Retrieved August 30, 2016, from http://paulbourke.net/miscellaneous/interpolation/

Braben, D., & Bell, I. (1984). *Elite*.

Dart, I. M., De Rossi, G., & Togelius, J. (2011). SpeedRock: Procedural Rocks Through Grammars and Evolution. In *Proceedings of the 2Nd International Workshop on Procedural Content Generation in Games* (p. 8:1–8:4). New York, NY, USA: ACM. 10.1145/2000919.2000927

Flood fill. (2016, May 29). In *Wikipedia, the free encyclopedia*. Retrieved from https://en.wikipedia.org/w/index.php?title=Flood_fill&oldid=722651188

Gustavson, S. (2005). Simplex noise demystified. *Linköping University, Linköping, Sweden, Research Report*.

Hello Games. (2016). *No Man's Sky*. Retrieved from http://www.no-mans-sky.com/

Hendrikx, M., Meijer, S., Van Der Velden, J., & Iosup, A. (2013). Procedural Content Generation for Games: A Survey. *ACM Trans. Multimedia Comput. Commun. Appl.*, *9*(1), 1:1–1:22. 10.1145/2422956.2422957

Interactive Data Visualization. (2009). *SpeedTreeRT*. Interactive Data Visualization.

Johnson, L., Yannakakis, G. N., & Togelius, J. (2010). Cellular Automata for Real-time Generation of Infinite Cave Levels. In *Proceedings of the 2010 Workshop on Procedural Content Generation in Games* (p. 10:1–10:4). New York, NY, USA: ACM. 10.1145/1814256.1814266

McMillen, E., & Himsl, F. (2011). *The Binding of Isaac*.

Mojang Synergies AB. (2009). *Minecraft*. Mojang Synergies AB. Retrieved from https://minecraft.net

Müller, P., Wonka, P., Haegler, S., Ulmer, A., & Van Gool, L. (2006). Procedural Modeling of Buildings. In *ACM SIGGRAPH 2006 Papers* (pp. 614–623). New York, NY, USA: ACM. 10.1145/1179352.1141931

Ong, T. J., Saunders, R., Keyser, J., & Leggett, J. J. (2005). Terrain Generation Using Genetic Algorithms. In *Proceedings of the 7th Annual Conference on Genetic and Evolutionary Computation* (pp. 1463–1470). New York, NY, USA: ACM. 10.1145/1068009.1068241

Perlin, K. (1985). An Image Synthesizer. In *Proceedings of the 12th Annual Conference on Computer Graphics and Interactive Techniques* (pp. 287–296). New York, NY, USA: ACM. 10.1145/325334.325247

Raffe, W. L., Zambetta, F., & Li, X. (2012). A survey of procedural terrain generation techniques using evolutionary algorithms. In *2012 IEEE Congress on Evolutionary Computation* (pp. 1–8). 10.1109/CEC.2012.6256610

Smelik, R. M., Tutenel, T., Kraker, K. J. D., & Bidarra, R. (2008). A Proposal for a Procedural Terrain Modelling Framework. In *ResearchGate*. Retrieved from https://www.researchgate.net/publication/267553398_A_Proposal_for_a_Procedural_Terrain_Modelling_Framework

Value Noise. (2016, July 31). In *Wikipedia, the free encyclopedia*. Retrieved from https://en.wikipedia.org/w/index.php?title=Value_noise&oldid=732399797

Author Biography

Jakob Schaal studied computer science at the Stuttgart Media University (HdM), Germany. Since 2012, he participated in the development of more than six computer games, many of which make heavy use of PCG algorithms. His passion for teaching and research is shown in his many online courses, some of which are free and some are directed at the premium market.

His main focus lies on PCG algorithms, for example, in the generation of levels for computer games. In the past years, he utilized many well-known algorithms like Perlin Noise for terrain generation, as well as developing completely new approaches which are able to fill empty 3D space with playable levels.

Schaal is a dedicated computer gamer. Since he was a child, he wondered how seemingly random yet interesting game levels are generated. His studies and findings are published in papers, books, and videos. He also participates in many open-source projects, some of which include PCG algorithms.

Procedural Adventure Generation: The Quest of Meeting Shifting Design Goals with Flexible Algorithms

Manuel Kerssemakers

Abstract This chapter presents the challenge of developing a procedural algorithm for creating exotic expeditions used in the best practice example *Renowned Explorers*, an adventure management game. A specific challenge was adapting the algorithm to shifting design requirements. To meet the goal, two methods for algorithm architecture are presented and compared. Firstly, static methods with a wide expressive range are employed, but ultimately fail. In consequence, requirements for a valid procedural design method are identified, resulting in a more flexible approach. The results of this work include not only a description of the algorithms but describe a new paradigm for creating procedural algorithms, which allow meeting shifting design goals.

1 Introduction

Since the beginning of video games, developers have tried to battle production costs and memory loads with the use of procedural content generation (PCG). However, there are more ways to extract value from PCG. Soon enough developers learned that procedural content was not only good for practical reasons, but could also provide creative advantages. A constant fresh challenge for the player, endless content and an element of surprise are all things a procedural algorithm can provide. In some games, the procedural algorithm is not simply responsible for filling in blanks of the world; it is responsible for generating interesting gameplay narratives. When these games are developed, the original game design and the supporting procedural algorithm can advance in no other way than in tandem.

In *Renowned Explorers* (2015), a modern adventure game presented as a best practice in this chapter, procedural content generation can offer many valuable

M. Kerssemakers (✉)
Abbey Games, Utrecht, The Netherlands
e-mail: manuel@abbeygames.com

© Springer International Publishing AG 2017

O. Korn and N. Lee (eds.), *Game Dynamics*,
DOI 10.1007/978-3-319-53088-8_9

things. The game allows the player to embark on *Indiana-Jones*-style adventures with their crew of three expert treasure hunters in a rich and optimistic world. To support the feeling of adventure and mystery, PCG is employed to provide the player with endless unexplored areas. These areas contain narratives and challenges and are built to represent a plausibly consistent area. The goal is to immerse the player in ancient deserts hiding great treasure, dangerous jungles brimming with smugglers and wildlife or the mysterious woods and hills of Transylvania.

PCG aims to create distinct artifacts that are both interesting and valid for the targeted use case. In the development of new games, this use case is hard to grasp, because as projects progress, games constantly change. Even with a proper pre-production, no one can guarantee that a better understanding of the game experience or a change in project scope does not result in changing design goals for whatever you're trying to generate.

In this chapter, we study the two challenges within the game development of *Renowned Explorers*. One of them investigates the actual algorithm used, and the other one the process of creating this algorithm.

1. Which algorithm can fulfill the requirements of high-quality adventure content in Renowned Explorers?
2. How can this algorithm be structured so that it can adapt, with proportional effort, to shifting design goals during the development cycle?

Supporting research questions are:

- What are the specific challenges of the content of the shipped version of *Renowned Explorers*, and how were they developed during development?
- Can a static but generic design succeed at efficiently supporting shifting requirements?
- What are desired properties of an algorithm architecture that can quickly adapt to unknown change?
- Is it tractable to work with a dynamic algorithm design?

To answer the research questions, we will start with examining the domain game of *Renowned Explorers* and the specifics of its requirements. As this chapter aims to explore not only the algorithms, but also the way in which they evolved during development, this section will also discuss changes to these requirements over time. In Sect. 2, we will look at state-of-the art research that informed the algorithm design. The middle section will consider a first paradigm in which two static but generic algorithms were created. These attempts try to capture a very broad so-called expressive range, which may contain all future possibly desired content. Section 4 explains why a pre-determined expressive space is unlikely to succeed and distills a number of requirements of algorithm design for PCG in developing original games. In Sect. 5, the final algorithm will be explained, while showing how its structure was flexible enough to adapt to changes in the game design. Finally, we will present conclusions to the research questions.

2 Requirement in the Best Practice Renowned Explorers

In this section, we discuss the requirements that the game of *Renowned Explorers* poses on the content it requires. This informs the procedural algorithm design, as it will become clear what output it needs to give. First, we will establish important qualities of the content. Then, we will review the history of requirements in the project, to understand how the PCG output had to change during development.

Renowned Explorers is a PC entertainment strategy game that lets you manage a team of explorers on their *Indiana-Jones*-like quest for treasure. The game has three layers of gameplay:

- a world-map layer, forming the highest level of abstraction
- an expedition layer in which the player explores locations
- a tactical arena layer which provides the lowest level of abstraction, challenging the player to solve conflicts.

Both the expedition layer and the arena layer work with procedurally generated space. Because the arena layer uses a subset of the techniques used by the expedition layer and poses a more traditional level generation challenge, we will focus on generating content for the expedition layer. For a glance at the arena generation, refer to Fig. 1. It shows an arena for which the movement graph is generated on a

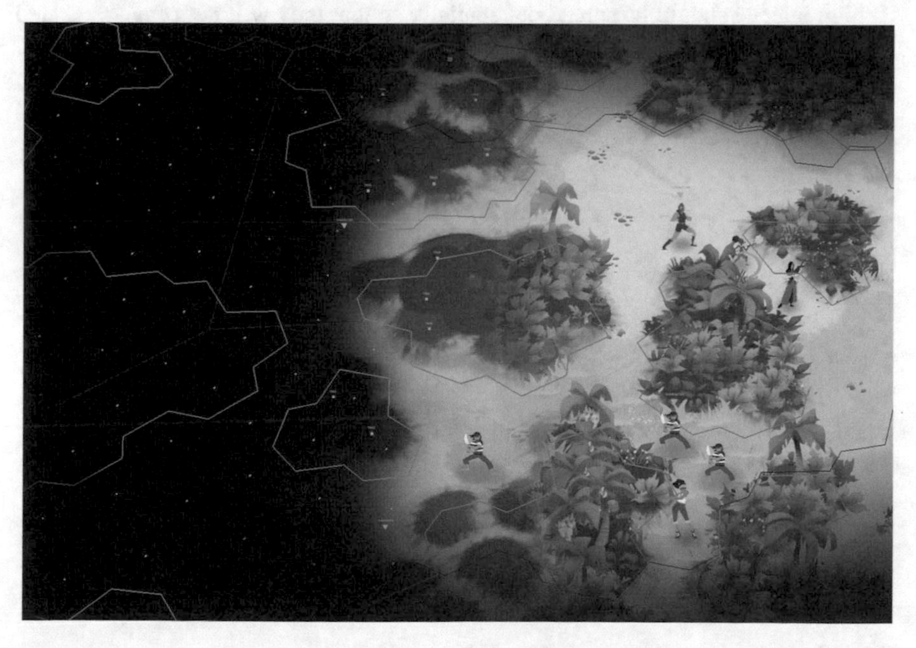

Fig. 1 An analysis of an arena as generated on a two-layered grid. Composed by an artist from the schematic and final views of the same level

refined tile grid, which in turn is generated on a coarse grid that represents slightly themed areas.

In the middle layer of the game, you navigate your team through exotic areas node by node, lifting the fog of war, encountering choose-your-own-adventure-style narrative stories (also called events or stories) on each one. To provide the player with interesting areas to explore on each new playthrough, we procedurally generate these exotic areas, also called expeditions. Both the navigation puzzle and the narrative content put requirements on this layer. This combination makes expedition generation a unique problem, for which an original algorithm is required (Fig. 2).

The requirements for such expeditions are as follows:

- **Narrative progression**. Each expedition has a main quest as input. The player needs to be introduced to the story and challenge at the start and conclude the expedition with a narrative conclusion. There are possibly narrative elements that need to be discovered before the conclusion can be reached.
- **Gameplay pacing**. The core loop of the expedition layers lets the player choose and travel to a node, and then plays out the story based on the player's choices. This is repeated until the expedition is either lost or completed. To enforce the pacing of this core loop, each node has a single story.
- **Narrative pacing**. Some nodes present more interesting and elaborate stories or challenges than others do. To present the player with a good pacing, points of high interest should be spread out spatially so that they will not often be visited consecutively.

Fig. 2 Example of a Caribbean expedition in Renowned Explorers

- **Narrative setting**. Each event may have setting constraints, requiring the story to be set on a subset of the available biomes (i.e., encountering monkeys only in the jungle and not on the beach).
- **Visual plausibility**. Expeditions need to support the game's theme by immersing the player into the location. To this end, the expedition's shape and structure need to plausibly convey the represented location such as an island or a desert. This requirement means that biome grouping is plausible such as beaches being close to the ocean. It also means that nodes are to some extent uniformly distributed over the expedition's space so the player's progress can be approximated by explored surface area and travel distance.
- **Navigation challenge**. The player has a limited number of supplies. Traversing edges costs between 0 and 3 supplies, usually 1. Each expedition needs to pose a non-trivial but solvable challenge of reaching the end before the supplies run out. This requirement poses constraints on adjacency, and path length between nodes of interest.
- **Event coexistence**. While in general each node contains an isolated story, they can be connected or even depend on each other. Some events specify the existence of other events as hard constraints.
- **Controlled randomness**. Events loosely belong to different classes such as "conflict," "skill check," "treasure" or "special/rare event." The expedition as a whole needs to guarantee a good mix of events as specified by the designer.
- **Configurability**. The same algorithm is used to generate a variety of exotic locations. As such, the algorithm should be configurable so that it will generate only expeditions fulfilling the specific requirements of that space. These requirements act as meta-requirements informing most of the above requirements.

Not every of these requirements existed in this form or at all at the beginning of the project. To understand the changes to which the algorithm design needed to adapt to, we shortly discuss the changes in these requirements during the development process.

The most radical change was the decoupling of the expedition and arena layer, which were initially one and the same. This change added story events to the expedition and removed opponents and tactics from the expedition layer, allowing it to focus more on storytelling and supply management. This change happened quite early in the project, but not before the first prototypes of the algorithm were developed (see approach 1 in the section on implementing static algorithms).

Over time, the expedition layer started to rely more and more on the events for creating a sense of place. Before the start of this change, each expedition needed to be visually unique and structurally distinct to create interesting difference. Throughout the change, location structures became more and more alike as we put emphasis on pacing, the navigation challenge and the relative location and coexistence of events. Main quests could initially be quite complex, asking the player to visit multiple locations or giving the player choices in a branching quest. Later on the design settled on a simple main quest, which only asked the player to reach a certain end node, leaving as much time as possible for player-driven exploration.

For a long and final period of iteration, the gameplay in the narrative events became ever more important. Consequently, more control was required on the part of the designer, emphasizing the requirements of event coexistence, controlled randomness and configurability. *Renowned Explorers*' expeditions turned out to be complex artifacts with many interlocking requirements that could not have been predicted beforehand by design.

3 State of the Art

3.1 Industrial Development Paradigms

It is no secret that procedural content generation plays an important role in today's games. Their use is widespread and diverse, and can be put on a spectrum of how tightly coupled the algorithm is to the game's design. On one end, there are content generation algorithms like *SpeedTree* (2009), which is used to fill in environments with foliage. Surely, the game art and feel is slightly influenced by the workings of the algorithm, but the main goal is preventing blanks, putting hardly any importance on the content itself.

On the other end, there are algorithms in games like *Spelunky* (2008), *Borderlands* (2009) or *No Man's Sky* (McKendrick 2015), where the algorithm and the game can hardly be viewed separately, as procedural aspects are at the core of the game design. At the extremities on this side, you will find PCG-based game design (Togelius et al. 2013). Depending on where a concrete application lies on that spectrum, algorithm design will differ. For the purpose of this chapter, we will look at the middle of the spectrum, where algorithms do service important design goals, but where design will not be compromised for the sake of the algorithm. Another way to put this is that procedurality itself is not part of the design and hand-crafted content would have worked at least as well, save for production limits.

3.1.1 Expressive Range

A useful vocabulary for talking about generators is presented by Smith (2012). Specifically, the work on defining and analyzing the expressive range (Smith and Whitehead 2010) of generators is relevant to this chapter. The expressive range is described as the style and variety of the levels that can be generated. For a designer, it is important to know the expressive range of a generator—expressed in a space which dimensions are the desired qualities. The designer does not design a single level, or the generator itself, instead they design the range of possible outcomes: the expressive range. When each generated artifact is important and being scrutinized by the player, the bar of quality is high. Several paradigms are interesting in this case.

3.1.2 Answer Set Programming

One of those is the use of answer set programming for procedural content generation (Smith and Michael 2011). This approach puts emphasis on the expressive range, stimulating designers to explicitly describe the expressive range, or "latent design space" as it is called in the paper. By writing a set of logical constraints, one can formally model the desired space. A domain-independent algorithm finds possible answer sets for the constraints, which are in turn interpreted as a piece of content, for example a level. This way the designer can iterate directly on sculpting the expressive range. This approach works well with puzzle-type games where logical constraints can easily model gameplay values.

3.1.3 Artist-in-a-Box

Another way of designing a PCG algorithm is described by Compton (2016). This is not limited to games, but aimed at computational creativity, the field in which we expect algorithms to come up with creative output. The approach is to try and design an artist-in-a-box, by understanding as well as possible how human designers would create a piece of content. The goal of the developer is to capture this process in an algorithm.

3.1.4 Mixed Initiative Design

A third approach to creating generators for gameplay is mixed-initiative design like in *Tanagra* (Smith et al. 2010) and *Refraction* (Butler et al. 2013). Just like answer set programming, this approach focuses on iteration. Firstly, a parameterizable generator with some arbitrary initial parameter set is created. A collection of created artifacts is shown to a user or designer who will rate or select among these. This serves as input for the algorithm, which will modify its parameters to center its output around the indicated bias. This approach is similar to the Procedural Procedural Level Generator Generator (Kerssemakers et al. 2012) where the core program is a generator of generators. The designer helps the external generator to find the inner generator that satisfies their needs. The danger is that the expressive range of the external generator does not contain a satisfying generator. In other words, the assumptions of the original generator may be wrong.

3.2 Types of Procedural Algorithms

3.2.1 Agent-Based Terrain Generation

At several stages, the PCG in *Renowned Explorers* drew inspiration from other algorithms. An example is agent-based terrain generation (Doran and Parberry 2010). These authors explore a way to generate terrain through configurable intelligent agents, operating on a heightmap. Each agent works on their own natural feature like hill formation, erosion or river creation. By configuring the amount of agents, their parameters and their order of execution, a designer has control on a feature level. As the agents work on the same terrain and are influenced by each other's work, this method leads to interesting emerging terrain features.

3.2.2 Voronoi

Level generation is often more easily solved in discrete space. We see this in the earliest dungeon crawlers, where games such as *Rogue* and *Nethack* use the grid of the monospaced fonts for its levels. Later on, there are different types of grids, like triangle or hexagonal grids. The hex grid proved to be popular with games as *Settlers of Catan* and *Civilization V*. All these examples use regular grids. This type of grid has some advantages:

- **Memory**. A regular grid is a repeating pattern, which makes saving its data in memory easy. For example, it is not required to keep adjacency data as the index of a rectangular tile (x, y) already indicates its neighboring tiles *f.e.* $(x + 1, y)$.
- **Uniform distribution**. For many gameplay purposes, a regular grid is useful. For example, it is easy to compute a distance between two points of a grid, as the distance in tiles is proportional to the distance in Euclidean space.

However, regularity is in no way a constraint for discretization of space. In games, we often encounter country maps using discretization. For gameplay reasons, the discretization of maps is frequently done without a regular grid. For a historical game like *Europa Universalis IV* (Europa Universalis 2013), it is important to work with countries as an entity, independent of their exact size.

One way of discretizing space is the Voronoi diagram as introduced in the nineties by Aurenhammer (1991). A Voronoi diagram is defined on a set of points and separates cells based on distance to each hub. Each cell consists of all the space that is closest to the corresponding point. In games, this can be used to discretize space around points of interest such as country capitals. One advantage is that Voronoi diagrams can have less or more local refinement as required. While they may look too chaotic to be used for games, these diagrams are only as "chaotic" as the point sets they are generated for. Any regular grid is a Voronoi diagram of the midpoints of its tiles. This means that Voronoi diagrams can represent a spectrum of discretizations starting at regular and extending toward chaos. This happens to be a

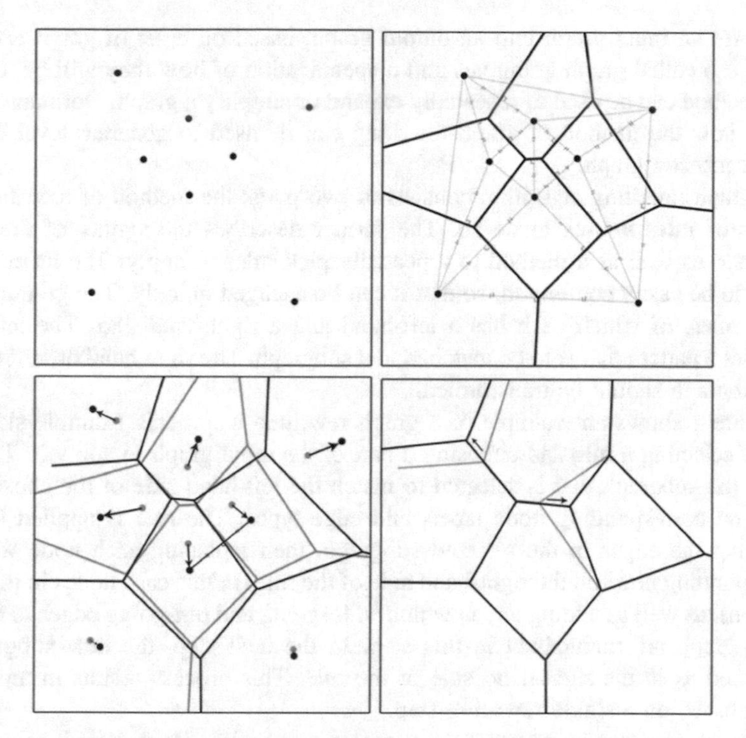

Fig. 3 A step-by-step irregular space discretization. *Top-left* shows original point set. *Top-right* shows the corresponding Voronoi diagram. *Bottom-left* shows Lloyd's algorithm being executed. *Bottom-right* shows a fix-up step to enforce a minimal side length

way to capture some desirable properties of regular grids and some of non-regular discretizations. A way of moving on this spectrum is to use Lloyd's algorithm (Lloyd 1982). This algorithm moves the centroid of each Voronoi cell and computes a new diagram from the new set of centroids. It can be iterated to make a Voronoi diagram more evenly spaced step by step.

An important disadvantage of Voronoi diagrams is that storing them into memory can be more troublesome. Fortunately there is a data structure called the "Double Connected Edge List" (DCEL) (de Berg 2008). It enables intuitive operations on the grid like splitting or joining cells, or adding new cells. Figure 3 shows how the described techniques can be used to create an irregular grid for games. We will call such grids "Voronoi grids."

3.2.3 Graph Transformation

As noted by Rozenberg (1997), "graph transformation is considered a fundamental programming paradigm where computation includes specification, programming, and implementation." Graph transformation (or graph rewriting) can be used to transform an input graph into an output graph, based on a set of graph rewriting rules (also called graph grammar) and a specification of how they will be applied. This method can be used to repeatedly expand or simplify a graph. Dormans (2011) shows how the method of graph rewriting can be used to generate level content from a mission graph.

A graph rewriting algorithm consists of two parts: the method of rewriting and the set of rules it uses to do so. The former describes the syntax of the graph grammar, as well as a method to repeatedly pick rules to apply. The latter part is meant to be easily configured, so that it can be changed quickly. The grammar is a set of rules, of which each has a left-hand and a right-hand part. The left hand specifies a pattern that is to be matched to a subgraph. The right hand describes how this subgraph should be transformed.

Figure 4 shows an example of a graph rewriting step. This example skips the step of selecting a rule and choosing a part of the input graph to apply it. The box shows the subgraph that is selected to match the left-hand side of the chosen rule based on corresponding node labels and edge types. The rule is applied by first removing the edges in the selected subgraph, then replacing each node with the corresponding node on the right-hand side of the rule (in this case nodes in the same position), as well as adding any new nodes. In-going and out-going edges to the rest of the graph are maintained in this step. In the next step, the new subgraph is connected as in the right-hand side of the rule. This process results in an output graph, based on a single rewriting step.

This process can be repeated to generate a complex graph based on a single graph grammar. By randomizing rule and subgraph selection, a single combination of graph grammar and input graph can create a wide range of output graphs. Note

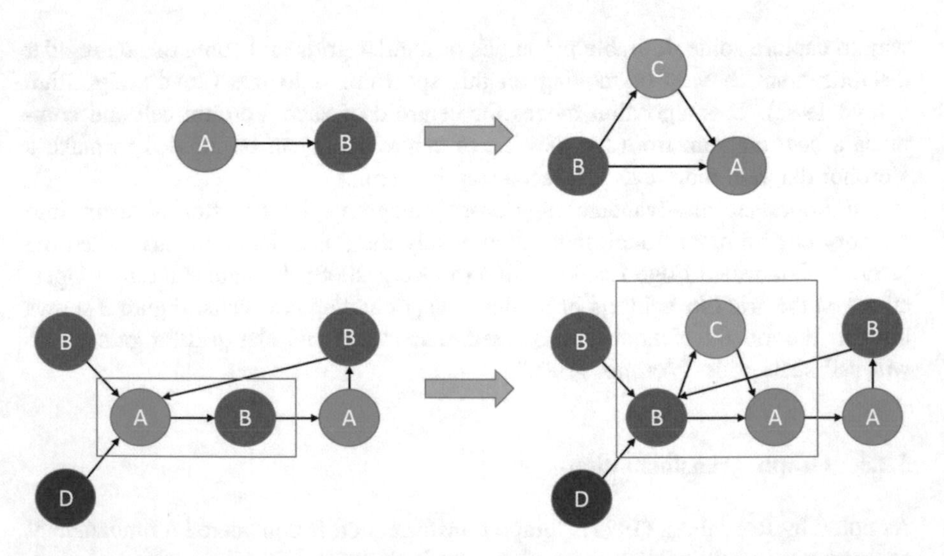

Fig. 4 A graph rewriting rule with a pattern on the *left-hand side* and a result subgraph on the *right-hand side*. *Bottom* An example of a graph in which the rule is applied to output a new graph

that configuring graph grammars does not allow exerting global control on output graphs, because its rules only enforce changes local to a certain step and subgraph during the rewriting.

In addition, graphs do not automatically represent space. If we treat each node as a small area in physical space and each edge as a fixed distance, it is easy to generate graphs that are impossible to map onto Euclidean space. Since game levels often need to be presented in such a space, this constitutes a problem. It can be solved by constraining the grammar to only allow physically plausible transformations, so that an input graph that can be mapped onto a certain space will always result in an output graph that can be mapped on the same space. Another solution is to layout the output graph after each rewrite step, possibly even reconnecting edges or deleting nodes to guarantee a physically plausible graph after each rewrite step.

4 Implementation of Static Generic Algorithms

All methods and paradigms described in the previous section informed the initial design of the procedural content generation of the best practice example of *Renowned Explorers*. However, the requirements had not yet settled in the early stages of development and the team was well aware of this. To give designers control over the PCG and avoid that programming would become a bottleneck during each iteration, we first investigated the possibility of having a relatively static algorithm which would not have to change in terms of architecture to meet future design requirements.

Instead, the designers would reconfigure the static algorithm whenever design goals changed, possibly requesting new features from the programming team. This method asked for a generic solution that could support any expressive range that might be targeted in the future. We investigated two possible solutions which will be presented in the following. The main question here was: Is it possible to build a static algorithm that can adapt with proportional effort, by means of configuration, to shifting design goals during development.

Approach 1: Graph rewriting
In line with answer set programming (ASP), we built a prototype algorithm that could support many expressive ranges, so that the designer would solely have to work on explicitly formulating the expressive range or latent design space. In the earliest stages of prototyping, all we knew about the space to move in was that it would be discrete, that each discrete element would have a biome attached to it, and that it needed to be shown on-screen in 2D Euclidean space.

Since we did not necessarily posit that space was uniformly distributed, we took the broadest expressive range thinkable to generate in, so we could encapsulate structural constraints in the configuration. This meant the space was nothing but a connected graph. The analogy with ASP would mean the designer describes the desired properties of the resulting graph in some formal language. Instead, the team chose to describe the graph by writing a graph grammar, describing the process of creating the content, rather than the desired content itself. This approach offered a more "natural" fit to capture our designers' level design knowledge, along the lines of the artist-in-a-box paradigm.

The resulting algorithm is described by the following steps:

1. Input is defined by the designer and consists of graph axiom A, grammar R as a set of graph rewriting rules with a fixed weight for each, the desired node count n and any custom conditions for the graph (in code). Note that A and R distinguish node types for later use for which the designer also inputs M, a mapping of node type to any gameplay content such as starting location, biome, obstacles or power-ups.
2. Start with a graph G as a copy of axiom A.
3. For each rule r in graph grammar R, list all the possible subgraphs of G that match the left-hand side of r.
4. Let H be the subset of R that contains only rules that have at least one match in G. If H is empty, terminate the algorithm.
5. Let rule r be a weighted random pick from H, based on fixed weights defined for each rule.
6. Apply r to a randomly selected subgraph match in G.
7. Repeat steps 1 through 6 until the algorithm terminates or reaches a set number of nodes. If G is a valid graph (i.e., has approximately the right number of nodes n and satisfies any custom conditions) accept, otherwise repeat the entire algorithm.
8. Gameplay content is allocated on basis of node types in G.

In this prototype algorithm, supply management and narrative were not yet a part of this game layer, so it had no way of dealing with these requirements. The algorithm resulted in expeditions like the one in Fig. 5.

Based on the prototype, we found that graph space was much larger than the targeted expressive range. We moved to a Voronoi-grid-based model (Fig. 6) to make the spatial distribution of cells workable. While in hindsight the graph representation was naïve, it is an example of not delineating an expressive range enough and trusting too much in later configuration to add structure to the output. This resulted in spending too much time on supporting unnecessary generic cases. Mapping arbitrary connected graphs onto Euclidean space with more or less fixed edge lengths and approximately uniform node distribution is a non-trivial problem, which implicitly became a development challenge.

Approach 2: Agent-based terrain generation
In a second attempt to capture a large set of expressive ranges, we developed another process-based configuration language. We worked with the new knowledge of approximately uniformly distributed tiles over space. In this stage, location structure was still dominant over the narrative requirements. We decided on an agent-based approach to assign biomes to tiles. This was inspired by Doran and Parberry's work (2010).

Fig. 5 An early prototype of Renowned Explorers with a graph-based expedition

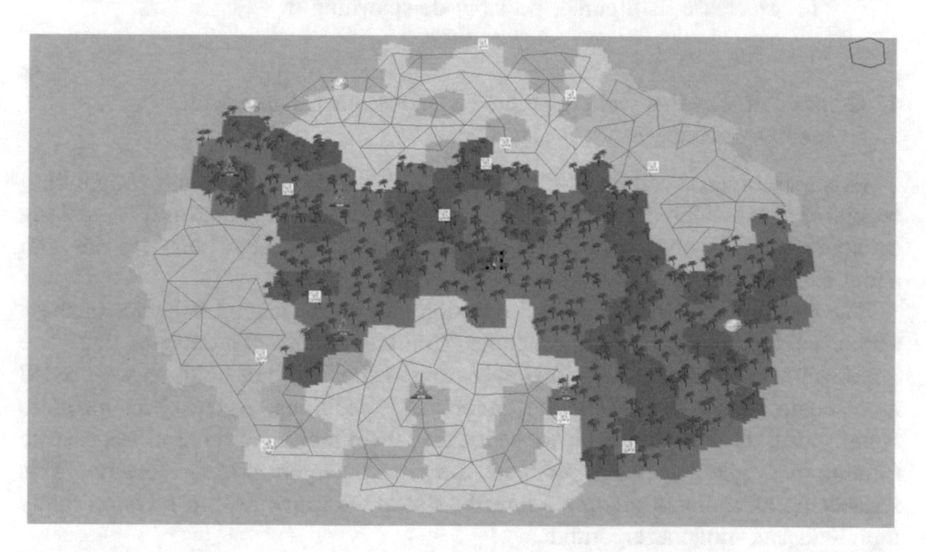

Fig. 6 Another early prototype, employing area generation based on a Voronoi grid

The basic idea was that many rules about how an expedition environment should look were local: Beach is adjacent to ocean, mountains are surrounded by plains, lakes are never very big, etc. The method was to configure

- software agent types that operate according to a specific randomized behavior and move on tiles close to them
- the number of agents per type, their operation time and resources, and their starting distribution over the Voronoi diagram.

With such a configuration for each expedition type, the agent's behavior could be run as a simulation. Randomness in the configuration implies a set of results, rather than a single outcome.

A typical prototype resulting from this approach is described as follows:

1. Input is given by the designer as: a set of agent types A with the behavior of each agent type a_i specified in code; a scheduling S which describes at which time steps agents of chosen types should be spawned. S contains information about the number of resources and steps that the agents have to operate and in which spatial pattern they should be spawned.
2. Initiate a timer $t = 0$.
3. Repeat until no agents exist
    a. Possibly spawn new agents as specified by S.
    b. For each agent
        i. Execute its *action*-code, which may result in a change to the tile biomed on their current position.
        ii. Execute its *move*-code, resulting in a new and adjacent position.
        iii. Decrease its lifetime, possibly de-spawning it.
    c. Increment t by 1.
4. Connect all tiles that were appointed a biome other than ocean.
5. Randomly place any story events.

Examples of agents are: land-creating agents turning ocean tiles into jungle tiles, beachline agents adding beach next to jungle tiles adjacent to ocean tiles, sinking agents that remove any type of land biome. Typically, the algorithm was run with about 6–20 agents over 20–100 time steps.

The resulting expeditions looked plausible due to the natural behavior of these agents and had wildly original structures, fulfilling several requirements that were important at the time (Fig. 7). However, this method did not hold, as we needed more control over features of the generated expeditions. Agents became more complicated. At the same time, we sequenced them more and more as opposed to running them parallel, which produces the most emergent original features. This sequencing of complex processes already shows the move toward a conceptually representative, modular algorithm.

To conclude: Can a static but generic design efficiently support shifting requirements? It might, but the designs we prototyped could not. In practice, the algorithms could never reach a state where they did not need new features to support newly originating design requirements. Had the effort to implement these features been proportional to the conceptual change, it could have been efficient. However, many new requirements pruned the targeted expressive range heavily. To have the algorithm create the desired output without losing generality would have cost significant effort, with a high chance of being unnecessary. This trend of needing more control could be specific to *Renowned Explorers* as a project.

Fig. 7 Prototype with agent-based terrain generation. Agents create original but hardly controllable structures. The *blue line* is from a later stage of quest placement, described in Step 3 of the resulting algorithm

However, our experience from other game design projects indicates that progress in design often implicates the need for more control, as designers frequently create new requirements.

5 Requirements for a Flexible Algorithm Design

The failure of two algorithm designs gave rise to one of this work's research questions: What are the desired properties of an algorithm architecture that can quickly adapt to unknown change? An important property can be distilled from the findings in the previous section. Both described designs failed because supporting a wide expressive range simply proved too expensive. Thus, it can be concluded that the desired design should target an expressive range that is as specific as possible. We will call this property *specificity*.

To find more properties, we will zoom in on the last part of our research question. What does it mean to "quickly adapt to unknown change"? It is hardly possible to implement any change quickly, because the complexity of the change will always be proportional to the complexity of the shifted requirement. However, proportionally small changes in the algorithm can achieve small changes in design goals for the output. To achieve this, the algorithm requires *modularity*. Isolation of modules in code ensures that when the rules of one design aspect change, only one module will have to change accordingly.

It is also helpful, to keep small shifts of design requirements proportional to the implicated change. This means is that when designers suggest a change, they will have an idea of the size of this change, based on their conceptual understanding of the algorithm. Therefore, the algorithm structure should mimic the designer's thought process. The artist-in-a-box analogy is very useful in this regard. Our goal then is to let the algorithm be a step-by-step reproduction of the designer's process. This results in the requirement of *sequentiality*. The agent-based terrain generation—approach 2 in the previous section—is an example of a parallel process, which made it hard to know how to improve the output by tweaking the algorithm.

While using a sequential structure, it is important to make no assumptions about what the sequence itself will be. A designer will make demands like: "Before the algorithm determines which cells are beach cells, it should mark which cells will contain boss stories." At the same time, the program may currently assume that environment generation and narrative distribution are separate steps. To avoid large refactors, the aim is to keep these generative systems modular, and let operations on them be sequential, but do not structure the systems to expect certain input. The "trick" is to keep the data for generated artifact separate from the generative systems. This is similar to the way the model and the controller are separated in the well-known Model-View-Controller design pattern (Krasner et al. 1988).

In summary, desired properties for our algorithm architecture are:

- specificity
- modularity
- sequentiality.

6 A Flexible Algorithm Design

In this section, we will answer the research question: Is it tractable to work with a dynamic algorithm design? It is answered by an example: a non-static algorithm design fulfilling the requirements described in the previous section. The resulting flexible procedural algorithm generating *Renowned Explorers'* expeditions uses exactly this design. To verify the approach, we will look at examples of how the algorithm design made it possible to adapt the algorithm itself with changes proportional to any shifting requirements.

Data structures

To address the requirements of modularity and global data access, data are structured as a set of objects with each one exposing an interface for parts of the algorithm to operate on. The static designs already had their model separated from the generation logic as well—the only difference here is that the model is subdivided into separate layers, which can reasonably be expected to be independent of each other:

- *Tile Grid*: The basic space we operate on, as generated by the Voronoi method.
- *Influence maps*: Simple instruments that generate fields of decreasing value around points or areas of interest (Uriarte and Ontañón 2012). By generating influence around existing points of interest for the player, we can easily determine which places on the map are still less interesting. This technique is simple and flexible. For example, we also use it to generate influence on the edges of the map, so that interesting narratives happen more often in the center of the map.
- *Biome Options*: Throughout the process, we track the biomes every tile (or cell) of the Voronoi diagram can still contain. At every step, some cells may already be defined as having a specific biome, while others still have multiple options. This object also keeps track of the amount of cells, which have to be placed for every biome. The way designers specify this, is by defining adjacency for every biome. These adjacency constraints can be hard constraints that force adjacency or soft constraints that stimulate or de-stimulate adjacency.
- *Event Placement*: This object tracks the placement of events (also called narratives) on specific tiles. Each tile can contain at maximum one event. We also keep track of any narrative dependencies inside this object. Dependencies are modeled as required events that will need to be placed in the future as soon as a depending event is placed.

In line with the requirement of sequentiality, the algorithm is designed to be executed in a step-by-step fashion.

Step 1. Grid Generation
The very first step of the algorithm is to start with a non-regular grid. To create a grid, a set of points is sampled. The sample method is a separate module. Initially, we used very chaotic patterns that only guaranteed some sort of distribution of points over space. The resulting Voronoi diagrams had many local characteristics that worked well to create visual differences between expeditions. Later on, a more reliable grid was required so we sampled points by taking the centroids of tiles in a virtual hexagon grid. These centroids were then perturbed before being fed into the Voronoi creation. Figure 8 shows the tile grids resulting from this grid generation method. In the final algorithm for the expedition generation, we used a grid that almost equals a hexagon grid, but looks slightly more organic by breaking the pattern with small perturbations. Ultimately, the design could probably have been realized with a hexagon grid, but the Voronoi approach made the algorithm much more flexible by providing a spectrum of grids to work with.

Step 2. Area Shape Generation
While some tiles are used for the expedition map, most of them will not. In order to find the tiles that make up the terrain, we first assign each tile a value by feeding its location into a heuristic module. Then, we determine how many tiles should be used based on the designer input, we call this amount n. We select the n tiles with the

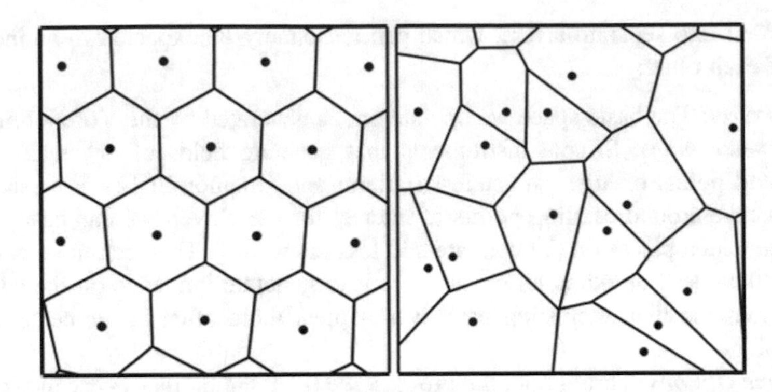

Fig. 8 Two Voronoi diagrams based on different point-sampling methods. On the *right* a chaotic pattern, on the *left* a slightly perturbed hexagon grid

highest values to use for the terrain. The writer of the heuristic function is responsible for keeping this set connected. If the set is indeed connected, this step also guarantees a connected movement graph by iteratively linking each tile to either the adjacent starting node or a node that has previously been connected. Depending on the designer input specifying the average number of connections per tile, extra connections are made and supply cost values are assigned.

The heuristic module allowed experimenting easily with many different shapes, albeit through code. The algorithm settled on egg-like shapes after trying various distorted circles, star-shapes and more complex composited shapes.

This step is an interesting example to illustrate shifting requirements. Not only did the heuristic allow easy shape changes, but also this entire step had a different context. One of the changes that took place in designing the algorithm is the order in which area shape and the main quest structure were determined. Originally, the main quest determined the shape of the area. This involved generating an abstract graph representation of the main quest, containing, for example, a start node and two or three different waypoints in some order. Later on, the design settled on a single fun main quest structure, which worked best on a certain island shape. This meant changing the algorithm so that it guaranteed a perfect shape beforehand. Since the main quest structure was the same every time, we could have changed the algorithm to result always in the correct shape based on that quest structure, still allowing for other quest structures. This would have been a more generic solution, but not the most time efficient, nor the most controllable solution. Design knew the exact required expedition shape, and the best way was to create a module generating this shape. This is an example of enforcing the specificity requirement.

Step 3. Main Quest Placement

The designer is allowed to input a main quest. This is a simple connected graph with a narrative attached to each node. As mentioned in the previous step, there was ultimately no variation in main quest structures. However, the main quest content can still define dependencies. Since the main quest is critical for the validity of the expedition, the algorithm places it before any other narratives.

Before the main quest settled on a single structure, it was used to generate terrain. This happened by overlaying the main quest graph on the Voronoi grid and claiming tiles within a certain distance of the graph.

Step 4. Place Special Narratives

When an island shape and a working main quest are assured, the next priority is to spread out interesting narratives over the space. To this purpose, we employ influence maps. An influence map is initialized, and influence is added around the main quest nodes and around the edge of the map. One by one, special "rare" events are placed on the least influenced points. These points reflect places that could use some excitement for the player. After each placement, the influence map is updated. The biome options are updated as well, because rare events may limit these.

In this phase, there are still many options for the algorithm and we do not expect the algorithm to fail placement. In the worst case, some design goals are stretched and points of interest spawn too close to each other. Some of these special events convert tiles around it by pre-determining custom biomes. This is an example of how the narrative heavily influences the environment. The implication is that shifting requirements in both narrative and environmental aspects of these important nodes resulted in required changes for the algorithm. The specificity, sequentiality and modularity helped greatly to iterate quickly on this step.

Step 5. Determine rest of biomes

At this point, the main features of an expedition are ready. What is left is filling the space with more small narratives and applying a plausible distribution of biomes. From experience with running the algorithm, the biggest risk was that a desired biome count could not be satisfied. For that reason, we first determinate any remaining ambiguous space to a certain biome. This process starts with the hardest biomes, picking tiles for biomes that have very little more options left than they need tiles placed.

Step 6. Place remaining narrative

After this step, the entire terrain is completed, with only the narrative margins left to fill in. In this phase, we have two last design goals: (1) Making sure existing narratives fulfill their dependencies. (2) Coupling biomes to interesting and relevant narratives.

We first place "twins," stories that are required by other stories. These stories can in turn have new dependencies, creating a recursive structure. We continue placing these twin events until there are none left. The designer is responsible for not overusing this feature, so that the algorithm will not run out of tiles to place narratives on. For remaining story-less cells, we make a controlled random draw from a pool of stories for the biome of the cell that need filling. Narratives implicating unplaceable dependencies are rejected.

Repeating this process eventually results in a completely filled expedition. The placement of new narratives is a complicated process, because there is no fail-safe for when dependencies cannot resolve. This means we have to analyze beforehand whether or not a narrative and its entire dependency tree can be placed, given the empty cells left, taking into account their biomes. This is the last step, as it finalizes the gameplay content of the expedition. Only after this phase, the mechanical model is turned into a visual representation employing height maps, texture blending, semi-random asset distribution and custom decorative art. This visual generation is beyond the scope of this chapter.

At each step, we chose the quickest-to-implement solution to solve the goal, making sure that the solution could be extracted or moved if so required at a later point. By not worrying about potential future use cases, we easily satisfied the requirement of specificity. Sequencing all the different steps fulfilled the modularity requirement. The use of heuristics further increased the modularity of the described algorithm. Thus, the algorithm satisfies all proposed requirements. It proved to be a flexible algorithm: It could adapt to shifts in requirements with effort proportional to the conceptual change.

Separating data from generative systems and obeying a sequential structure is a way to visualize intermediate results of your algorithm. It is even possible to keep a history of executed steps and roll-back to or re-execute any step of your procedural generation. It was only when we started visualizing intermediate results that *Renowned Explorers'* level design progress was no longer bottlenecked by programming.

The static generic algorithms failed, because their expressive range was much too large and design could not be expressed simply by configuring the algorithm, rather than changing it. This does not mean that there is no role for configuration in a flexible algorithm. Configuration is applied not to adapt the algorithm to shifting requirements, but to delineate parts of its expressive range. By configuring the algorithm differently for each location the player visits, the procedural generator can be used to create many different types of content.

7 Conclusion

Expedition generation in the game *Renowned Explorers* posed both spatial and narrative requirements that needed to be addressed by a procedural algorithm. However, the main challenge of creating procedural algorithms for in-development game designs lies in structuring an algorithm so that it can adapt to shifting requirements. In this best practice example, we presented the path toward the design of procedural algorithms that need to adapt to shifting design goals.

We compared two ways of designing algorithms: One approach is static but generic algorithms that have a large expressive range and may therefore in their output contain the desired latent output. Two attempts were presented: a method based on graph rewriting, and another one employing terrain-transforming software agents. Both approaches did not hold. The static algorithms failed because supporting their wide range of configurations puts an expensive overhead on algorithm changes. Without these changes, the algorithm and its configuration remained static, which increased the difficulty for designers to control it in their preferred manner.

From the failure of the static approach, we distilled requirements for a more flexible approach in algorithm design:

- *Specificity* of the expressive range makes sure implementation never costs more work than is required for the desired output.
- *Modularity* keeps the algorithm maintainable, as its components might need to be swapped out, reordered or removed.
- *Sequentiality* stimulates an analogy with the designer's process. Thus, required algorithm changes are prone to mimic the proposed conceptual changes. It also helps visualizing the procedural process intuitively.

Based on these requirements, we presented the development of a flexible algorithm design for *Renowned Explorer*'s expeditions, which was maintained throughout development. The procedural algorithm keeps track of the expedition model, which can be visualized to ultimately deliver a finished gameplay artifact. As design goals changed, it was possible to maintain and develop the flexible algorithm with effort proportional to the conceptual changes.

Next to the algorithm for creating adventures in *Renowned Explorers*, the results of the work presented here is a new paradigm for developing future PCG algorithms: Flexible algorithm design is well suited for all games where design and procedural content generation are interlocked.

References

Aurenhammer, F. (1991). Voronoi diagrams—a survey of a fundamental geometric data structure. *ACM Computing Surveys (CSUR)*. Retrieved from http://dl.acm.org/citation.cfm?id=116880

Borderlands. (2009). Retrieved from https://en.wikipedia.org/wiki/Borderlands_(video_game)

Butler, E., Smith, A. M., Liu, Y.-E., & Popovic, Z. (2013). A mixed-initiative tool for designing level progressions in games. In *Proceedings of the 26th annual ACM symposium on User interface software and technology* (pp. 377–386). ACM.

Compton, K. (2016, February 22). So you want to build a generator. Retrieved from http://galaxykate0.tumblr.com/post/139774965871/so-you-want-to-build-a-generator

de Berg, M. (2008). *Computational Geometry: Algorithms and Applications*. Springer.

Doran, J., & Parberry, I. (2010). Controlled procedural terrain generation using software agents. *IEEE Transactions on Computational Intelligence in AI and Games*, 2(2), 111–119.

Dormans, J. (2011). Level Design As Model Transformation: A Strategy for Automated Content Generation. In *Proceedings of the 2Nd International Workshop on Procedural Content Generation in Games* (pp. 2:1–2:8). New York, NY, USA: ACM.

Europa Universalis. (2013, August 13). Retrieved from http://www.europauniversalis4.com/

Kerssemakers, M., Tuxen, J., Togelius, J., & Yannakakis, G. N. (2012). A procedural procedural level generator generator. In *2012 IEEE Conference on Computational Intelligence and Games (CIG)* (pp. 335–341).

Krasner, G. E., Pope, S. T., & Others. (1988). A description of the model-view-controller user interface paradigm in the smalltalk-80 system. *Journal of Object Oriented Programming*, *1*(3), 26–49.

Lloyd, S. (1982). Least squares quantization in PCM. *IEEE Transactions on Information Theory/ Professional Technical Group on Information Theory*, *28*(2), 129–137.

McKendrick, I. (2015, July 21). *Building A Galaxy: Procedural Content Generation in No Man's Sky*. Presented at the Nucl.ai, Vienna. Retrieved from https://archives.nucl.ai/recording/building-a-galaxy-procedural-generation-in-no-mans-sky/

Renowned Explorers. (2015, September 2). Retrieved from http://www.renownedexplorers.com

Rozenberg, G. (1997). *Handbook of Graph Grammars and Comp*. World Scientific.

Smith, A. M., & Michael, M. (2011). Answer Set Programming for Procedural Content Generation: A Design Space Approach. *IEEE Transactions on Computational Intelligence in AI and Games*, *3*(3), 187–200.

Smith, G. M. (2012). Expressive design tools: Procedural content generation for game designers. Retrieved from https://escholarship.org/uc/item/0fn558gq.pdf

Smith, G., & Whitehead, J. (2010). Analyzing the Expressive Range of a Level Generator. In *Proceedings of the 2010 Workshop on Procedural Content Generation in Games* (pp. 4:1–4:7). New York, NY, USA: ACM.

Smith, G., Whitehead, J., & Mateas, M. (2010). Tanagra: A Mixed-initiative Level Design Tool. In *Proceedings of the Fifth International Conference on the Foundations of Digital Games* (pp. 209–216). New York, NY, USA: ACM.

SpeedTree. (2009). Retrieved from http://www.speedtree.com/

Spelunky. (2008, December 21). Retrieved from http://www.spelunkyworld.com/

Togelius, J., Champandard, A. J., Lanzi, P. L., Mateas, M., Paiva, A., Preuss, M., & Stanley, K. O. (2013). Procedural content generation: Goals, challenges and actionable steps. *Dagstuhl Follow-Ups*, *6*. Retrieved from http://drops.dagstuhl.de/opus/volltexte/2013/4336/

Uriarte, A., & Ontañón, S. (2012). Kiting in RTS Games Using Influence Maps. In *Eighth Artificial Intelligence and Interactive Digital Entertainment Conference*. Retrieved from http://www.aaai.org/ocs/index.php/AIIDE/AIIDE12/paper/viewPaper/5497

Author Biography

Manuel Kerssemakers is co-founder of Dutch entertainment game studio Abbey Games, which to date has released two successful core strategy game titles on PC. His main function is gameplay and AI programmer, but as a founder, he also has responsibilities in design, marketing, and business development. Manuel also appears internationally as a public speaker about game programming and marketing.

In 2007, he started his studies with the bachelor cognitive artificial intelligence at the Utrecht University. After graduating cum laude with a thesis about Co-evolution in Genetic Programming and Tron, he pursued the Game and Media Technology master. His focus was on simulations, games, and procedural generation. This led to a semester at the IT University of Copenhagen where he worked on automatic balancing of asymmetric Real-Time-Strategy game unit sets and created a "Procedural Procedural Level Generator Generator."

Kerssemakers grew up loving games with a living simulation inside of it such as Settlers 2, Black & White or Rollercoaster Tycoon. During high school and university, he created his own games but it was not until he met the other three founders of Abbey Games that he started creating games of interesting scope.

In 2012, the studio released *Reus*, a 2D PC god game. In 2015, they released *Renowned Explorers*, an adventure management game. Kerssemakers also regularly participates in game jams, where he won several awards. His goal has always been to include intelligent behavior in games, either by implementing original AI or by programming algorithms that can make clever or creative choices on their own.

Index

A
Agent-based terrain generation, 156, 160, 163, 164
AI Director, 3, 9
AltspaceVR, 96
America's Army, 48
Answer set programming (ASP), 155, 159
Artist-in-a-box, 155, 160, 164
Audiosurf, 7

B
Battlefield 4, 47, 48
Bestiarium Invocation Toy, 122
Bézier surfaces, 116
Binding of Isaac, The, 132
Blend shapes, 100, 119, 121, 123, 125, 130
Borderlands, 8, 10, 11, 20, 23, 48, 154
Brownian motion, 3, 4
Business Model Canvas, 31

C
Call of Duty, 3, 20
Character design, 33, 95, 115, 129
Choose-your-own-adventure, 152
Cityengine, 23
Civilization V, 156
Color cycling, 126
Color palettes, 126, 128
Competition, 71, 72, 74, 88
Cooperation, 71, 72, 74

D
Design, Dynamics, Experience (DDE) framework, 27, 42
Design, Play and Experience (DPE) framework, 30
Destiny, 100
Diablo, 17, 22, 23
Diablo II, 8, 126
Diablo III, 132
3D meshes, 115
Doodle Jump, 10
Dragon Age, 100

E
Elemental Tetrad, 29, 30, 33
Elder Scrolls, The, 96, 100, 104
Elder Scrolls IV, The : Oblivion, 8, 24
Elite, 8, 18, 21, 23, 132
Europa Universalis IV, 156
Expressive range, 149, 150, 154, 155, 159, 160, 162, 163, 168, 169

F
Facial symmetry, 98, 99
Fallout, 96, 104
Flow state, 10
Fractals, 3, 4, 6, 11, 21

G
Game bits, 8, 19
Gameful Design process, 31
Game space, 8, 19
Gamification, 10, 31
Gamification Model Canvas, 31
Gang Beasts, 117
Golden ratio, 99, 100
Gradient mapping, 127

© Springer International Publishing AG 2017
O. Korn and N. Lee (eds.), *Game Dynamics*,
DOI 10.1007/978-3-319-53088-8

Graph grammars, 158, 160
Graph rewriting, 157–160, 168
Graph transformation, 157
Gunshot sounds, 47–49, 53, 56, 58–60, 64–66

H
Height map, 131, 135, 136, 142, 143, 168
Human average face, 102, 103

I
Implicit surfaces, 117
Impossible Creatures, 116, 123
Influence maps, 165, 167
Irregular grids, 157

K
.Kkrieger, 8, 18, 19

L
Latent design space, 155, 159
Lathe sculpting, 119, 120
Left 4 Dead, 9, 10
Lloyd's algorithm/Lloyd's relaxation, 157
Ludo-narrative Disconnect, 41

M
Mandelbrot sets, 3–5, 11
Marching cubes, 117
Markov chains, 3
Mass Effect, 96, 100, 104
MDA framework, 27–29, 31, 32, 35, 36, 39, 42
Mechanics, Technology, Dynamics, Aesthetics and a Narratives Framework (MTDA+N), 29
Metaballs, 117, 125
Minecraft, 2, 8, 10, 23, 132
Mixed initiative design, 155
Model-View-Controller, 164
Modular characters, 117, 118, 122, 129
Modularity, 163, 164, 167–169
Mortal Kombat, 126
Motivation, 1, 31, 72–74, 76, 88
Motivation Design process, 31

N
Nethack, 21, 156
No Man's Sky, 117, 130, 132, 154

O
Octalysis gamification framework, 31
Ontogenetic procedural generation, 115, 116

P
Palette switching, 126
Paper's Please, 118
Parametric synthesis, 50
Perlin noise, 6, 11, 131, 132
Player-Centered Design, 31
Player modeling, 73, 89
Player-Subject, 35–39, 41, 42
Point sampling, 166
Procedural audio, 47
Procedural content generation (PCG), 1, 15, 16, 18, 19, 23, 47, 149, 154, 155, 159, 169
Procedural level generation, 155
Procedural Procedural Level Generator Generator, 155

R
Refraction, 155
Renowned Explorers, 149–152, 154, 156, 159, 161, 162, 164, 168, 169
Rigblocks, 117
Rogue, 8, 17, 20–23, 156

S
Self-avoiding random walk, 4, 5
Sentinel, The, 21
Sequentiality, 164, 165, 167, 169
Settlers of Catan, 156
Shooters, 11, 18, 20, 48, 60, 129
Simplex noise, 6, 131, 132
Sims, The, 96, 100, 104
Social facilitation, 74, 87
Sony Sports Champions, 73
Sound synthesis, 47, 48, 57, 60, 61
Specificity, 163, 164, 166–169
SpeedTree, 24, 145, 154
Spelunky, 8, 23, 154
Spore, 117, 123, 128–130
Stack Gun Heroes, 48
S.T.A.L.K.E.R.
Shadow of Chernobyl, The, 24
Super Mario Bros., 126

T
Tanagra, 155
Taxonomy, 15, 16, 18, 19, 21
Teleological procedural generation, 116
Tiny Wings, 10

U
Uncanny valley, 116
User experience, 11, 29, 41
UV mapping, 123

V
Value noise, 131–138, 141, 145, 146
Voronoi diagram, 142, 156, 157, 161, 165, 166

W
Wiener Process, 4
Witcher 2, The, 24
World of Warcraft, 96, 97, 100

X
X, a game of YZ, 119

Printed in the United States
By Bookmasters